ADSL

Standards, Implementation, and Architecture

The CRC Press Advanced and Emerging Communications Technologies Series

Series Editor-in-Chief: Saba Zamir

Data and Telecommunications Dictionary, Julie K. Petersen

Handbook of Sonet Technology and Applications, Steven S. Gorshe

The Telecommunications Network Management Handbook, Shervin Erfani

Handbook of Communications Technologies: The Next Decade, Rafael Osso

ADSL: Standards, Implementation, and Architecture, Charles K. Summers

Protocols for Secure Electronic Commerce, Ahmed Sehrouchni and Mostafa Hashem Sherif

After the Y2K Fireworks: Business and Technology Strategies, Bhuvan Unhelkar

Web-Based Systems and Network Management, Kornel Terplan

Intranet Management, Kornel Terplan

Multi-Domain Communication Management Systems, Alex Galis

Mobile Intelligent Agents Applied to Communication Management Systems, Alex Galis and Stefan Covaci

ADSL
Standards, Implementation, and Architecture

Charles K. Summers

Series Editor-in-Chief
Saba Zamir

CRC Press
Boca Raton   London   New York   Washington, D.C.

Library of Congress Cataloging-in-Publication Data

Summers, Charles K
   ADSL: standards, implementation, and architecture/ Charles K. Summers
      p. cm. — (Advanced and emerging communications techniques)
   Includes bibliographical references and index.
   ISBN 0-8493-9595-X
    1. Data transmission systems. 2. Telecommunications—Standards. 3. Modems. 4. Telephone switching systems, Electronic. I. Title. II. Series.
   TK5105.S86 1999
   621.38—dc21

                                                      99-26897
                                                        CIP

    This book contains information obtained from authentic and highly regarded sources. Reprinted material is quoted with permission, and sources are indicated. A wide variety of references are listed. Reasonable efforts have been made to publish reliable data and information, but the author and the publisher cannot assume responsibility for the validity of all materials or for the consequences of their use.
    Neither this book nor any part may be reproduced or transmitted in any form or by any means, electronic or mechanical, including photocopying, microfilming, and recording, or by any information storage or retrieval system, without prior permission in writing from the publisher.
    The consent of CRC Press LLC does not extend to copying for general distribution, for promotion, for creating new works, or for resale. Specific permission must be obtained in writing from CRC Press LLC for such copying.
    Direct all inquiries to CRC Press LLC, 2000 Corporate Blvd., N.W., Boca Raton, Florida 33431.

    Trademark Notice: Product or corporate names may be trademarks or registered trademarks, and are only used for identification and explanation, without intent to infringe.

© 1999 by CRC Press LLC

No claim to original U.S. Government works
International Standard Book Number 0-8493-9595-X
Library of Congress Card Number 99-26897
Printed in the United States of America 1 2 3 4 5 6 7 8 9 0
Printed on acid-free paper

For my beloved wife Marie,
children Cheyenne, Michael, and Jonathan,
and
friends and family

All texts extracted from International Telecommunication Union (ITU) documents by the author of this book have been reproduced with the prior authorization of the ITU as copyright holder. The sole responsibility for selecting extracts for reproduction lies with the author and can in no way be attributed to the ITU. The complete volume(s) of the ITU material, from which the texts reproduced are extracted, can be obtained from:

International Telecommunication Union
Sales and Marketing Service
Place des Nations
CH-1211 Geneva 20 (Switzerland)

Telephone: + 41 22 730 61 41 (English)
+ 41 22 730 61 42 (French)
+ 41 22 730 61 43 (Spanish)
Telex: 421 000 uit ch
Fax: + 41 22 730 51 94
X.400: S = Sales; A=400net; P = itu; C = ch
E-mail: Sales@itu.int
http://www.itu.int/publications

All texts extracted from American National Standards Institute (ANSI) documents by the author of this book have been reproduced with the prior authorization of the Alliance for Telecommunications Industry Solutions (ATIS) as copyright holder. The sole responsibility for selecting extracts for reproduction lies with the author and can in no way be attributed to ATIS or ANSI. The complete volume(s) of the ANSI material, from which the texts reproduced are extracted, can be obtained from

ANSI Sales Department
ANSI
11 West 42nd Street, 13th Floor
New York, New York 10036
Tel: +1 212 642 4900

Table of Contents

Acknowledgments xi
Acronyms and Abbreviations xiii
Introduction xvii

1 Analog and Digital Communication
1.1 Communication Forms 1
    1.1.1 Analog 2
    1.1.2 Digital Transmission Coding 3
1.2 Transmission Media 5
    1.2.1 Copper Wiring 6
    1.2.2 Other Transmission Media 6
1.3 Switching and Routing 7
    1.3.1 Basics of Switching 8
    1.3.2 Circuit-switches and Packet-switches 9
    1.3.3 Routers 11
        1.3.3.1 LANs and WANs 11
        1.3.3.2 Functions of the Router 12
1.4 Multiplexing 12
1.5 Infrastructure Limits 13
    1.5.1 Distance Limitations on Local Loops 15
    1.5.2 Loading Coils 15
    1.5.3 Repeaters, Amplifiers, and Line Extenders 16
    1.5.4 Bridged Taps 16
    1.5.5 Digital Loop Carriers (DLCs) 16
    1.5.6 Summary 17
1.6 Bottlenecks 17
    1.6.1 Host I/O Capacity 17
    1.6.2 Access Line Capacity 18
    1.6.3 Long-distance Line Capacity 19
    1.6.4 Network Saturation 19
    1.6.5 Server Access Line and Performance 19
    1.6.6 Summary 19

2 The xDSL Family of Protocols
2.1 From Digital to Analog 21
2.2 Digital Modems 22
2.3 The ITU-T, ADSL, and ISDN 23
2.4 ADSL Standardization 24
    2.4.1 Standards Bodies 25

  2.4.2 ADSL Standards Bodies..25
    2.4.2.1 ADSL Forum and UAWG....................................26
    2.4.2.2 ANSI..26
    2.4.2.3 ETSI...27
    2.4.2.4 ITU-T...27
2.5 The xDSL Family of Protocols..27
  2.5.1 56K Modems...31
  2.5.2 BRI ISDN (DSL)..31
    2.5.2.1 Physical Layer...32
    2.5.2.2 Switching Protocol...35
    2.5.2.3 Data Protocols...37
  2.5.3 IDSL..37
  2.5.4 HDSL/HDSL2...38
    2.5.4.1 Signaling Using Channel Associated Signaling.................40
    2.5.4.2 Signaling Using Primary Rate Interface ISDN...................41
    2.5.4.3 HDSL2 or SHDSL...41
  2.5.5 SDSL...42
  2.5.6 ADSL/RADSL..42
  2.5.7 CDSL/ADSL "Lite"..43
  2.5.8 VDSL..45
2.6 Summary of the xDSL Family...45

3 The ADSL Physical Layer Protocol
3.1 CAP/QAM..47
3.2 Discrete Multitone..49
3.3 ANSI T1.413...50
  3.3.1 Bearer Channels...51
  3.3.2 ADSL Superframe Structure..55
    3.3.2.1 Fast Data and Interleaved Data...............................57
    3.3.2.2 Fast Byte..58
    3.3.2.3 Sync Byte and sc Bits..58
    3.3.2.4 Indicator Bits..58
    3.3.2.5 CRC bits..60
  3.3.3 Embedded Operations Control...60
3.4 ADSL "Lite"...62
3.5 ATU-R Versus ATU-C..64
3.6 DSLAM Components...64

4 Architectural Components for Implementation
4.1 Open Systems Interconnection Model (OSI)...67
  4.1.1 Layer 1 (Physical Layer)..68
  4.1.2 Layer 2 (Data Link Layer)...69
  4.1.3 Layer 3 (Network Layer)...70
  4.1.4 Layer 4 (Transport Layer)..71
  4.1.5 Upper Layers...71

Books are to be returned on or before
the last date below.

	4.1.6	Interlayer Primitives	72
	4.1.7	Protocol Modularity	72
4.2	Hardware Components and Interactions		73
	4.2.1	Interface Chip	74
	4.2.2	Physical Layer Semiconductors	75
	4.2.3	System Configuration Design	75
		4.2.3.1 Host-controlled Systems	76
		4.2.3.2 Coprocessor Systems	77
		4.2.3.3 Standalone Systems	77
4.3	Protocol Stack Considerations		77
	4.3.1	Signaling	77
	4.3.2	Interworking	78
	4.3.3	Stack Combinations	78
4.4	Application Access		79
	4.4.1	Host Access	79
	4.4.2	Control Systems	80

5 Hardware Access and Interactions

5.1	Semiconductor Access		82
	5.1.1	Memory Maps	84
	5.1.2	I/O Requests	84
	5.1.3	Registers	84
	5.1.4	Indirect Register Access	85
	5.1.5	Data Movement	85
	5.1.5.1 FIFOs		85
	5.1.5.2 Buffer Descriptors		86
5.2	Low-Level Drivers		87
	5.2.1	Primitive Interfaces	88
	5.2.2	Interrupt Servicing and Command Handling	88
	5.2.3	Synchronous and Asynchronous Messages	88
5.3	State Machines		89
	5.3.1	States	89
	5.3.2	Events	90
	5.3.3	Actions	91
	5.3.4	State Machine Specifications	91
	5.3.5	Methods of Implementation	92
	5.3.6	Example of a Simple State Machine	92
5.4	ADSL Chipset Interface Example		94

6 Signaling, Routing, and Connectivity

6.1	Signaling Methods		98
	6.1.1	Analog Devices	98
	6.1.2	Channel Associated Signaling (CAS)	101
	6.1.3	Q.921/Q.931 Variants	101
6.2	Routing Methods		102

	6.2.1	Internet Protocol	103
	6.2.2	Permanent Virtual Circuits	104
		6.2.2.1 ATM Cells	104
		6.2.2.2 Frame Relay	105
6.3	Signaling Within the DSLAM		105

7 ATM Over ADSL
7.1	B-ISDN (ATM) History, Specifications, and Bearer Services	108
	7.1.1 Broadband Bearer Services	108
	7.1.2 Specific Interactive and Distribution Services	109
7.2	B-ISDN OSI Layers	110
7.3	ATM Physical Layer	111
7.4	ATM Layer	111
	7.4.1 ATM Cell Formats	113
	7.4.2 Virtual Paths and VIrtual Channels	115
7.5	ATM Adaptation Layer	116
	7.5.1 AAL Type 1	118
	7.5.2 AAL Type 5	118
7.6	ATM Signaling	120
	7.6.1 Lower Layer Access	120
	7.6.2 General Signaling Architecture	120
	7.6.2.1 User-side States	121
	7.6.2.2 Network-side States	122
	7.6.3 B-ISDN Message Set	123
	7.6.4 Information Elements	125
7.7	Summary of ATM Signaling	126
7.8	System Network Architecture Group (SNAG)	126

8 Frame Relay, TCP/IP, and Proprietary Protocols
8.1	Frame Relay	129
	8.1.1 Frame Relay Data Link Layer	130
	8.1.2 Link Access Protocol for Frame Relay	131
	8.1.2.1 Address Field	131
	8.1.2.2 Congestion Control	133
	8.1.2.3 Control Field	134
	8.1.3 Data Link Core Primitives	134
	8.1.4 Network Layer Signaling for Frame Relay	136
	8.1.5 MultiProtocol Over Frame Relay	137
8.2	Internet Protocol	138
	8.2.1 The Data Link Layer	138
	8.2.2 IP Datagrams	139
8.3	Transmission Control Protocol	141
	8.3.1 TCP Virtual Circuits	142
	8.3.2 TCP Header Fields	142
	8.3.3 TCP Features	144

8.4	Proprietary Protocol Requirements	144
	8.4.1 Data Integrity	144
	8.4.2 Data Identification	145
	8.4.3 Data Acknowledgment	145
	8.4.4 Data Recovery	146
	8.4.5 Data Protocol	146

9 Host Access

9.1	Ethernet	148
	9.1.1 History	149
	9.1.2 OSI Model Layer Equivalents	149
	9.1.3 The Medium Access Control (MAC)	150
	9.1.4 The Ethernet Frame	152
	9.1.5 Physical Medium and Protocols	154
	9.1.6 MAC Bridges	254
9.2	Universal Serial Bus	155
	9.2.1 Goals of the USB	155
	9.2.2 USB Architecture	156
9.3	Motherboard Support	157
	9.3.1 Data Bus Extension	157
	9.3.2 Microprocessor Direct Access	158

10 Architectural Issues and Other Concerns

10.1	Multi-Protocol Stacks	160
	10.1.1 Architectural Choices	160
	10.1.2 Software Implementation	161
	10.1.2.1 "Physical Layer" Replacement	162
	10.1.2.2 Coordination Tasks	163
	10.1.2.3 Data Structure Use	164
10.2	Signaling	165
10.3	Standardization	165
10.4	Real-Time Issues	166
	10.4.1 Bottlenecks	166
10.5	Migration Needs and Strategies	167
	10.5.1 Replacement of Long-Distance Infrastructure	168
	10.5.2 FTTN, FTTC, and VDSL	168
10.6	Summary of Issues and Options	170

References and Selected Bibliography 173
ITU-Rcommendations 173
Other Technical References 174
Selected Internet Uniform Resource Locators (URLs) 175
Selected Bibliography 175

Index 177

Acknowledgments

First, I would like to thank Gerald T. Papke, former editor at McGraw-Hill and CRC Press, who persuaded me to write this book. Thanks also go to Dawn Mesa at CRC Press for her patience while I juggled family life, work at TeleSoft International, and writing this book. Thanks go to other editors and writers who, over the years, have helped me to work toward creating better books. Any errors still remaining are solely my responsibility.

I would also like to acknowledge the various people in my life that made this book possible. Many thanks to my beloved wife, Marie, who made time in our lives for me to write this book, acted as encourager, and worked as an extra proofreader. Next, thanks go to Charles D. Crowe, my business partner and friend, and all the other employees of our company TeleSoft International, Inc. Thanks also go to Cheryl Eslinger of Motorola and Kathleen Gawel of Capital Relations, Inc. for their help with the Motorola CopperGold™ API. Finally, I would like to thank Palma Cassara of GlobeSpan Semiconductor, Inc. for information useful in better understanding CAP (and other) ADSL products.

And since this is a book about computer technology, I would also like to "thank" the machines and programs that made it possible: to Apple Computer for my Power Macintosh™ G3 and for AppleWorks™ 5.0, to Hewlett-Packard for my LaserJet™ 5M, and to Corel® for continuing to support WordPerfect™.

Acronyms and Abbreviations

AAL	ATM Adaptation Layer
ADSL	Asymmetric Digital Subscriber Line
ANSI	American National Standards Institute
AOC	ADSL Overhead Channel
AS0-3	ATM downstream simplex sub-channel designators
ASx	One of the ATM downstream sub-channels
ATM	Asynchronous Transfer Mode
ATU-C	ADSL Transceiver Unit, Central office end
ATU-R	ADSL Transceiver Unit, Remote end
B-ISDN	Broadband ISDN
BECN	Backward Explicit Congestion Notification
BRI	Basic Rate Interface
C-plane	Control plane
CAP	Carrierless Amplitude and Phase modulation
CAS	Channel Associated Signaling
CCITT	International Telegraph and Telephony Consultative Committee (old name of ITU-T)
CLP	Cell Loss Priority
CO	Central Office
CODEC	COder-DECoder
CPE	Customer Premise Equipment
CRC	Cyclic Redundancy Check
CRV	Call Reference Value
CS	Convergence Sublayer
CSA	Carrier Serving Area
DE	Discard Eligibility
DLCI	Data Link Connection Identifier
DMT	Discrete MultiTone
DTE	Data Terminal Equipment
DTMF	DualTone MultiFrequency
EOC	Embedded Operations Channel
ETSI	European Telecommunications Standards Institute
FCS	Frame Check Sequence (HDLC)
FDM	Frequency Division Multiplexing
FEC	Forward Error Correction
FECN	Forward Explicit Congestion Notification (Frame Relay)
FEXT	Far-End crossTalk
FTTC	Fiber To The Curb
FTTN	Fiber To The Neighborhood

HDLC	High-level Data Link Control
HDSL	High-speed Digital Subscriber Line
HDTV	High Definition TeleVision
IDSL	ISDN Digital Subscriber Line
IE	Information Element
IEEE	Institute of Electrical and Electronics Engineers
IETF	Internet Engineering Task Force
IP	Internet Protocol
ISDN	Integrated Services Digital Network
ISO	International Organization for Standardization
ITU	International Telecommunication Union
ITU-T	ITU-Telecommunication Standardization Sector (formerly called CCITT)
LAN	Local Area Network
LAPB	Link Access Procedure Balanced
LAPD	Link Access Procedure on the D-channel
LAPF	Link Access Procedure Frame-mode
LAPM	Link Access Procedure for Modem
LLD	Low-Level Driver
LMI	Local Management Interface
LS0-2	Duplex sub-channel designators
LSB	Least Significant Bit
MAC	Media Access Control
MODEM	MOdulator-DEModulator
MPEG	Motion Picture Experts Group
MSB	Most Significant Bit
N-ISDN	Narrowband ISDN
NEXT	Near-End crossTalk
NIC	Network Interface Card
NNI	Network-Network Interface
NT	Network Termination
NT1	Network Termination 1
NT2	Network Termination 2
OAM	Operation And Maintenance
OSI	Open Systems Interconnection
PABX	Public Access Branch Exchange
PBX	Private Branch Exchange
PCM	Pulse Code Modulation
PDU	Protocol Data Unit
PLP	Packet Layer Protocol
POH	Path OverHead
POTS	Plain Old Telephone System
PPP	Point-to-Point Protocol
PRI	Primary Rate Interface
PSTN	Public-Switched Telephone Network
PVC	Permanent Virtual Circuit

QAM	Quadrature Amplitude Modulation
QOS	Quality Of Service
RAM	Random Access Memory
RFC	Request For Comment
ROM	Read-Only Memory
SAR	Segmentation and Reassembly
SDH	Synchronous Digital Hierarchy
SDL	Specification Description Language
SM	Service Module
SONET	Synchronous Optical Network
SSCOP	Service-Specific Connection-oriented Protocol
STM	Synchronous Transfer Mode
SVC	Switched Virtual Circuit
TC	Transmission Convergence sublayer
TCP	Transmission Control Protocol
TE1	Terminal Equipment 1 (ISDN)
TE2	Terminal Equipment 2 (non-ISDN)
TEI	Terminal Endpoint Identifier
TDM	Time Division Multiplexing
U-plane	User plane
UART	Universal Asynchronous Receiver/Transmitter
UNI	User-Network Interface
USB	Universal Serial Bus
VCC	Virtual Channel Connection
VCI	Virtual Channel Identifier
VC	Virtual Circuit
VPC	Virtual Path Connection
VPI	Virtual Path Identifier
WAN	Wide Area Network

Introduction

Asymmetric Digtal Subscriber Line (ADSL) use is one of the general Digital Subscriber Line (xDSL) techniques. While it has been around in the laboratory for about ten years, this particular technique has since shifted to the special evaluation site to the beginnings of consumer access. By the time this book is available, some mass provision of ADSL to the general consumer market will be available.

Digital Subscriber Line is just that—use of digital transmission methods on the carrier line that commonly exists between a local switching location and the home subscriber. Arguments can be made that xDSL, by definition, includes the common modems that have been in use for the past 20 years, as well as new techniques such as cable modems which make use of subscriber lines—but not the *same* subscriber lines as are used by ADSL and its close relatives.

Most definitions, however, include only the techniques used over the ubiquitous lines that have been used for Plain Old Telephone Service (POTS) over the past century. This definition limits the number of protocols to be considered, as well as ensuring that the limitations that have entered into the telephone network are taken into account with the use of the newer methods. If new lines, including fiber optics, are used for new services then the physical plant (wiring, connections, junctures, etc.) can be architected for the most optimum use with the service.

The existing twisted-pair copper wiring exists worldwide as part of the gradually constructed infrastructure used to support speech communication. Since this slowly expanding system has developed over the past 100 years, it is not surprising that the needs of speech have been the main criteria of network design. This has helped to improve the quality of speech services over the network and allowed interpersonal communication on a global basis.

Communications techniques are always changing—primarily to be able to communicate faster and over greater distances. Using a system in the same way for 100 years might now be considered to be a long time, however, previous systems lasted many hundreds, even thousands of years. Today we are faced with steadily decreasing cycles of time where the needs of the network will have greatly different requirements.

This doesn't mean that the old communication techniques will simply disappear. People will still talk, write, telegraph, and use "regular" speech phone service. The same is true about the infrastructures that are put into effect to support those services. It is not economically (or, in some ways, socially or politically) possible to yank out all of the old wiring and replace it with the current "best" method or replace the old equipment with new.

So, the new techniques must coexist with the old and leverage the ability to make use of the existing structures to support the new. It is within this context that we will examine xDSL and ADSL.

The existing switched network was engineered specifically for use in supporting speech communication. The development of facsimile (fax) machines to make use of the same network for graphic data transmission didn't change the general criteria too much. Modem use, however, did make a difference by changing the duration of average calls. Still, this was not a significant difference as only a relatively small percentage of people did lengthy Bulletin Board System (BBS) or other electronic message system access.

The big danger, indicating potential overwhelming of the existing switched networks, arose out of speed and multiple-access mechanisms such as the Internet. A 1200-bits per second (bps) modem takes so long time to transfer data that physical transfer via express shipping companies continued to be a very competitive choice. At 38,600 bits per second, however, transfer times start to make electronic distribution (for relatively small files) economically practical and this means that the speech network's traffic distribution criteria starts to go awry. 56K Modems and Base Rate Integrated Services Digital Network (ISDN) shift the formula more and more. The result is "brown-outs" where transmission systems are overwhelmed and line busy signals become more frequent.

The dilemma becomes how to make use of the existing (and very difficult and expensive to replace) infrastructure without causing these massive problems. The solution is to use the part that is the most difficult to replace and use new parts in the areas where it is more feasible. ADSL attempts to do this by utilizing the existing wiring between the home, or business, and the switching network and avoiding the existing network used for making speech calls.

The first item, therefore, is to make use of the existing twisted-pair copper wiring. This line (consisting of the wire and all equipment on the wire) has been engineered to efficiently support high-quality speech transmission. Some of these design criteria directly affect the ability to carry other types of data over the same wires. These conflicting criteria, and other difficulties in using the existing lines for new services, will be examined in Chapter 1 of this book.

In Chapter 2 we will discuss the various methods that can be used to make faster high-quality use of existing wiring. Earlier, we mentioned 56K modems and Basic Rate ISDN. The architecture of ISDN will be discussed in greater depth as well as the existing standards organizations and the various types of Digital Subscription Line (xDSL) transmission methods.

Chapter 3 deals with the specific physical transmission needs of ADSL. Since ADSL was invented in the laboratory, it has been necessary to conduct "trials" of different ADSL configurations and equipment to consider "real-life" infrastructure situations. These trials have helped to make equipment available for network and user equipment. It is unusual for equipment to "disappear" once it has been developed. This leaves us with new "legacy" equipment and other equipment which is in the winner's circle (agrees with the developed international standards). They will all continue to exist, at least for the time being, as new equipment evolves from laboratory experiment to everyday application.

Placing a new physical protocol on existing wires is only one step in new service capability. Equipment must be produced to support the protocol on *both* ends of the wire. This means that software and hardware must be created to work together.

Although the existing network is circumvented with the use of ADSL, the ability to connect to something else—end-to-end connectivity, must be there. Finally, the user must have access to the data in a way that they can use it productively. These issues are introduced in Chapter 4.

Hardware access is the topic of Chapter 5. In theory, it is possible to do any type of physical, or logical, protocol with a general microprocessor and the ability to control the physical characteristics of the signal. In practice, it is neither economical nor practical to do physical layer transmission in this way. Instead, specialized semiconductor chips are designed to allow data access without microprocessor concerns over specific physical line content. Low-Level Drivers (LLDs) allow the higher-level protocols to control the semiconductor devices.

Signaling, or the control of how the network makes connections, is the introductory topic in Chapter 6. The main areas that are considered are cell and frame relay, although some comparisons are made to the existing circuit-switched systems that are used in speech networks.

Asynchronous Transfer Mode (ATM), a form of Broadband ISDN, and cell relay switches are covered in Chapter 7. Cells are small units of data that can be switched rapidly on an individual basis. ATM allows these cells to be used as a set of data. As part of this, a set of signaling protocols have been defined to direct the cell relay network to set up connections on a semi-permanent or transient basis. Finally, the recommendations of the Service Network Architecture Group (SNAG) concerning the use of ATM (and PPP) over ADSL are discussed.

Frame relay is similar to ATM except that the frames are generally much larger than the cells. This lowers overhead but increases the size, and quantity, of buffers needed for practical routing of the frames. Transport Control Protocol/Internet Protocol (TCP/IP) is the underlying network control protocol used within the Internet. Since the Internet is one of the strong driving factors for development of higher speed connectivity, it makes sense for TCP/IP to be part of any discussion about possible architectures. A discussion of various proprietary methods of connecting ADSL endpoints to services completes Chapter 8.

An ADSL service has now been set up. The equipment has access to data at up to (perhaps) 8,000,000 bits per second. How is this transferred to the processors/applications that will make proper use of it? This is discussed in Chapter 9. Possible data transfer ports include older methods such as Ethernet, newer standards such as the Universal Serial Bus (USB), protocol-specific methods such as ATM-25, and the potential redesign of the motherboards on general purpose computers to allow direct access to ADSL (or other protocol) ports.

In the final chapter, Chapter 10, we bring together all aspects of ADSL use as they concern software architecture issues. These include assembling multiple-layer protocol stacks,—"nesting" one protocol within another; coordinating signaling control with data processes; examining special real-time issues dealing with protocol stacks; and, in closing, a look at migration strategies to ADSL and beyond.

As a collection of topics, one leading to the next, this book will endeavor to explain why and how ADSL will take its place within the family of data transmission protocols used around the world.

1 Analog and Digital Communication

Communication is the process of sending and receiving information. In the non-computer world, it is the process of providing information in a form that others can understand. It may be via voice, sound signals (drums, music, alert sounds, etc.), writing, sign language, body language, flashing lights, (or smoke signals), or something else. It is communication, however, only if someone else can understand. Voice (or sound signals) will not work to communicate with someone else if they don't know the meaning of the signals or if they are physically unable to capture the information.

The same situation occurs in the computer world. The process of communication is broken down into the tasks of transmission and reception. Similar to the non-computer world, both sides must be able to make use of the same physical medium. The physical medium is manipulated into signals and both ends (or, with broadcast signals, multiple-receiving ends) must know the meaning of the signals being used.

We therefore have a situation where there are two parts that must be compatible in order to communicate: physical and coding. The physical part refers to the medium and the coding pertains to how the medium is manipulated in order to make recognizable signals. A third level is protocol which is how the signals that are used are understood.

An analogy to the non-computer world can be made with speech. Sound waves are the basis of the physical medium. The codes are based on how those sound waves are changed. This might be in degrees of loudness, pitch, sub-tones, and so forth. The "protocol" would be a language (i.e., English). The protocol has two aspects which, in non-computer terms, may be called grammar and context. Grammar says that the symbols are formed correctly and context says that the symbols are *used* correctly. In the computer world, these aspects of protocols are considered to be *syntax* and *semantics* (the same can be used for human languages in a formal study).

The first part of this chapter will discuss the possible physical layers and the coding mechanisms available. It will then proceed to discuss those elements that make the medium more useful and easier or cause difficulties.

1.1 COMMUNICATION FORMS

A communications signal can take many forms depending on the medium used. If fiber optics are used then the medium will be light. Radio waves can be used or infrared waves can be used for short distances. Most of the documentation, however, addresses electrical transmission media, since this is the most prevalent form found in residential and business use. Even Fiber To The Curb (FTTC) often does not have the last lap as something other than electrical.

It is, therefore, reasonable to limit the discussion to electrical forms, and that will be the primary focus of this book. Most transmission media have two categories of signaling: analog and digital. As we will see, in the electrical transmission world, both are continuous signals. The difference is in the method of imposing signal meanings on the medium.

1.1.1 ANALOG

Analog signals are a continuous form with an infinite number of possible values. This is similar to that of sound, which in theory can take on any strength (amplitude) and pitch (frequency). This can be seen in Figure 1.1. Although the signal can take any of an infinite number of values, the equipment may not be able to produce, or receive or understand, all possible values. The human ear cannot perceive sounds of less than a certain volume or greater than another volume (although this range will vary from person to person). Similarly, the ability to create and receive different frequencies varies from person to person (and even more between species).

The first forms of electrical communication occurred in a very simple form: "off" or "on" coupled with duration. Morse code was developed to take advantage of this simple signaling form (see Figure 1.2). A "dot" was an "on" with a short duration. A "dash" was an "on" with a longer duration. The "off" was a period when the current was not applied. The signal was not necessarily continuous, and (today) it could certainly be considered to be digital as we will see in the next section.

However, the next signaling form to be widely used *was* continuous—the transmission of sound via electricity. By the use of mechanical components very similar in form and function to the human ear, the signal form was translated from audible

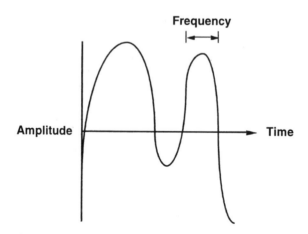

FIGURE 1.1 Analog speech/electrical example.

```
-  ---   --- - -- --
```
A   D   S   L

FIGURE 1.2 Morse code as an example of digital signaling.

to electrical, giving a signal that, once again, looked very similar to that shown in Figure 1.1 except that the change in signal occurred via current or voltage manipulations.

Carrying a potentially infinite number of signals is a great advantage, but transmission media have some common problems. They are the problems of degradation and attenuation. Degradation means that the signal loses its form. This usually occurs because of interaction with other signals of a similar nature. For example, a voice in a crowd will rapidly merge with those of other people and, at a certain distance, will be unintelligible. An electrical signal carried over wire, that is within a bundle, will be affected by other signals from other wires. It will also be affected by the imperfection of the medium—flaws in the wire and insulation.

Attenuation is associated with power. An analog signal is created at a certain point in time and space. As it moves from the point of origination (once again, moving either in time or space), the strength of the signal will fade as it gets further from the originating point of creation.

As we will see in the section on infrastructure limits, both degradation and attenuation can be managed by recreating the signal. However, analog forms, with their potentially infinite number of signals, are more difficult to recreate correctly and can only be recreated within certain tolerance levels. As the number of times that the signal is recreated increases, the chance of significant compounded errors (errors that are problems with recreating signals that have already had errors introduced) also increases.

So, analog transmission forms have the strength of being able to carry potentially infinite numbers of signals, but the problems of degradation and attenuation cause this strength to become a liability for the transmission of complex data requiring a low error rate. This leads us to a greater discussion of the second category of signal types: digital.

1.1.2 DIGITAL TRANSMISSION CODING

As mentioned above, the first form of electrical transmission may be considered to be digital. Digital means able to be counted (often considered to be on one's "digits," implying a base 10 scenario). Binary is the simplest form of digital coding—on or off, high or low. The main difference is that the signals are discrete; specific values from a fixed set are passed rather than a continuous set of potentially unlimited values. More generically, digital information consists of a set of limited values which vary at a fixed rate.

In theory, digital values are disjoint, as can be seen from the sample Morse code digital signal in Figure 1.2. When using electrical transmission media, however, it is better to use alternating voltages to reduce power consumption. This means two things: the "ideal" coding scheme would have an average electrical level of "neutral" and the variance will actually be continuous.

Figure 1.3 shows a more "real-life" electrical digital signal. Note that this signal form is continuous. It is also designed so that it can convey an infinite number of signal values. In the electrical transmission world, there is no explicit difference between the analog and digital forms; the difference lies in how the signal forms are used.

A continuous electrical signal is used digitally by the process of *sampling*. The signal is sampled, or tested, at precise time intervals. This value is interpreted according to a set of criteria, called the transmission code. For many simple transmission codes, this amounts to being a number of ranges. A value of +/-0.5 volts to +/-1.5 volts, for example, may be interpreted as the value 1, while a value between -0.5 and +0.5 volts is interpreted as the value 0. The actual differences in subvalues (such as between -0.4 and -0.3 volts) are ignored. This converts the continuous (potentially analog) form into digital.

Both negative and positive voltage levels were used in the above coding scheme. This is done for electrical reasons, to save power on the line (and to help prevent steadily increasing distortion as the physical medium is changed by the continued voltage) it is "ideal" if the average voltage is close to 0. This can be done by balancing the sample codes over the positive and negative ranges of potential values.

The sampling interval is also called the *clock rate*. The clock rate determines the amount of data that can be transferred over a period of time. The faster the clock rate, the greater the amount of data transferred in the time period. However, as the intervals decrease, it starts to approximate continuous analog signal interpretations and the potential error rate increases. Nyquest's Sampling Theorem states that the information transfer rate can only be 1/2 the speed of the sampling rate. In other words, if you want to send 10 data values per second, the data source must be sampled 20 times per second. Two sequential samples with the same interpreted value is considered usable. If the sampling does not have two identical values in a

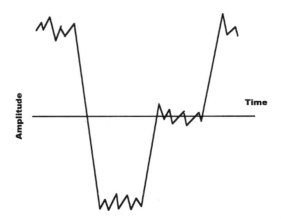

FIGURE 1.3 Digitally interpreted continuous signal.

Analog and Digital Communication

row, it means that the physical transmission is fluctuating in an illegal pattern and no usable data can be obtained.

The above example has the signal interpreted as possessing one of two possible values. It is certainly possible for there to be four potential values (or five, or nineteen). Because of standardization of digital computers on the binary data form, most coding schemes will involve values in powers of two (2, 4, 8, 16). Some example coding schemes can be found in Figure 1.4.

It is also possible to treat the electrical signal in a three-dimensional manner. While the above scenario has two dimensions, voltage and time, it is possible to have three dimensions: voltage, time, and phase. This allows for much greater information transfer rates with a wider separation of interpreted values. This is one of the methods used within ADSL coding schemes.

Note that, in both the two- and three-dimensional coding schemes, it is necessary to have a baseline against which to compare values. With a two-dimensional voltage scheme, the value of 0 is a natural baseline; with a three-dimensional method, either an explicit baseline form must be sent along with the coded signal or an implicit (such as the value 0 for two-dimensional schemes) must be used.

1.2 TRANSMISSION MEDIA

As stated earlier, copper wiring used for electrical transmissions will be the primary focus of physical level discussions. However, in a long-distance network, many different media are likely to be involved in transmission. A brief discussion of the various transmission media follows.

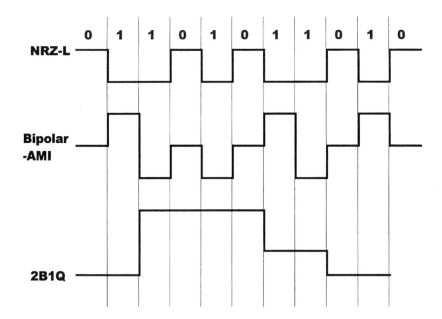

FIGURE 1.4 Examples of digital codes.

1.2.1 COPPER WIRING

Some of the early wiring for electrical transmissions made use of metals other than copper (e.g., iron and steel). It was soon determined that copper served as a good mixture of capabilities and cost-effectiveness. Copper has good conductivity and is sufficiently malleable to be able to be formed into wires of different sizes, bent, cut, and shaped into needed configurations. Gold provides an even better medium (and is thus used to a great extent for electrical connections in critical areas such as within electronic parts), but is cost-prohibitive for extensive use.

The first wires were simple single strands. However, when bundled with other wires, the signals tended to interfere with one another (called *crosstalk*). Using two wires as a pair and then twisting them together improved the resistance to crosstalk and also improved attenuation characteristics. Coating the wires before twisting further enhanced performance. To prevent each twisted pair from interfering with other pairs in a bundle, it would have been further useful to shield the pairs from each other but this was not done for standard wiring as it added to the expense.

The thickness of the wire is usually specified in North America according to the American Wire Gauge (AWG) standard. These numbers are basically reciprocals of diameter units so a thickness of 0.03589 (about 1/28) inches (0.9 mm) is called gauge 19, 0.02535 (about 1/39) inches (0.63 mm) is called gauge 22, and so forth. A higher number indicates a smaller diameter. The international metric community uses a direct metric measurement for standard wire sizes. Note that using wires of different thicknesses will change the electrical characteristics of the wire and using different thicknesses on the same line may cause problems. Generally, a thicker line will be able to carry a clearer signal for longer distances (but will cost more per foot/meter).

This unshielded twisted pair thus evolved as the primary medium used for building the international infrastructure for electrical communications. As is true for most developments of this kind, it was a result of sufficient technical capability with a cost low enough to be marketable. The important point is that it was devised with a specific set of criteria and those criteria have changed over the years. Using the old infrastructure within a new framework poses problems for both the developer and the manufacturer.

1.2.2 OTHER TRANSMISSION MEDIA

Transmission media are devised in accordance to the changing needs of the environment. They may be economic or technical (though the actual research may be largely theoretical and done for curiosity or challenge). Most of the time, the physical medium (or signaling methods imposed thereon) is devised, tested, improved, and then manufactured when it meets market needs. This was true of ADSL, which was devised as a research project within various research laboratories, including Bellcore.

Fiber optics are often considered to be the "best" medium to use with the current technology. If the current infrastructure were not already in place, it would likely be the medium of choice for ground-based transmission systems. In order to reach this point, it was necessary to solve a number of problems and have the ability to

Analog and Digital Communication

use supportive technologies with it. The laser was needed to provide sufficiently controllable light to provide signaling methods. In the early days of using fiber optics, methods of joining one fiber to another were very difficult (and therefore expensive). This had to be solved. Currently, fiber optics are cheaper to install and maintain, and provide a medium which supports greater speed than copper. However, ripping out the existing copper lines to residences and businesses "just" to replace it with fiber optics is not cost-effective. Many new long-distance lines (trunks) are being configured with fiber optics.

Regardless of how good fiber optics may be as a physical medium, they still require a continuous line between endpoints. It may be practical to put a line between Paris and Berlin or even between New York and London (submerging the line at the bottom of the Atlantic Ocean), but it isn't practical to have a line between Denver and the moon; nor is it the most cost-effective. Making use of satellite transponders, or microwave towers, to relay signals over difficult physical obstructions such as mountain ranges may be more useful.

Broadcast media, such as microwave transmissions, eliminate the need for a continuous link between the transmission and reception points. The "tighter" frequencies are easier to direct and control and suffer less from attenuation. However, since all signals commingle in the same physical area, there is a limitation to how many transmission "lines" can be in the same area.

This is why microwave and radio wave transmissions are regulated in terms of frequencies and power output. It would otherwise be impossible to distinguish between signal sources as they might overlap other sources. A radio transmitter may have a frequency of 530 KHz and an effective range (based on power) of 50 miles (80 km). With these limitations, it is permissible to have another station at a distance of 150 miles (240 km) to have the same frequency and power rating and not overlap. However, if they both had a range of 100 miles (160 km), there would be a region where receivers would be getting two separate signals on the same frequency, causing interference and making the signal unintelligible.

On the other hand, Personal Communication Systems (PCS) takes advantage of range limitations very effectively. By having roaming areas that are severely limited in range, it is possible to make use of a wide frequency range (spectrum) without significant interference from other devices. When the transmitter goes out of range from one area, the signal is picked up by another device. This is a hybrid method where the link is *not* continuous, but still provides uninterrupted transmission services (actually, disruptions do occur frequently, but the transfer period from one receiver to another is sufficiently short so they usually go unnoticed).

1.3 SWITCHING AND ROUTING

Given the fact that it is impractical to use the broadcast medium for all transmissions, it is necessary to ensure that the appropriate endpoints are connected. This connection is called a *circuit*. The endpoints form a circuit; the path along which the physical connection exists is called a *route*.

Theoretically, it would be possible to have all endpoints directly connected to one another. In a set of five endpoints, this would require 10 distinct lines (as shown in Figure 1.5) to allow each to have a connection to all the others. However, this progresses with the number of endpoints. To connect 10 endpoints directly to each other, 44 lines are needed. Obviously, this is impractical when the endpoints reach into the hundreds, thousands, or millions.

1.3.1 BASICS OF SWITCHING

The method of connecting the endpoints together without dedicated lines is called *switching*. This is accomplished by making the final connections only when needed. The first switches were human-operated "switchboards." Every subscriber had a line from their location to a central location; support of 10 locations required 10 lines. At the central office, the attendant was given the name of the party they wanted connected and the two lines were *bridged* together (using a "patch cord"). There now was a direct connection between the two endpoints. A switchboard of this type was practical for hundreds of lines. It would even be possible to "conference" more than two endpoints together at the central location.

The technology of switches has changed over the years. The last switchboard in the U.S. was retired in the late 1970s. In rural areas, there are still many "cross-connect" switches which provide an electro-mechanical method of "patching" connections together. However, most switching (and probably all long-distance switching) is now provided by some form of "electronic switch"— basically, a computer that is specialized to connect endpoints together.

For long-distance service, long-distance "trunks" were used. Although trunk lines are considered to be large-capacity connections, it is not an absolute requisite. Let's say for example, that one central office controlled 1,000 endpoints. If a subscriber wanted to talk to someone who was serviced at a different central office it would require two connections to be made—if a line existed directly between the

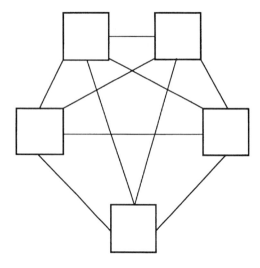

FIGURE 1.5 Full connectivity.

Analog and Digital Communication   9

central offices. Subscriber A would have a line to Central Office (CO) 1. This would be patched to the line from Central Office 1 to CO 2. At CO 2, the line would be connected to the line for Subscriber B. Note that it would require 1,000 lines between CO 1 to CO 2 to allow all of the subscribers at one CO to talk to all of the other subscribers at CO 2.

This additional set of connections, as is true for direct connections at any time, becomes impractical as the size of the *network* increases. So, what is done is that *traffic statistics* are taken. This might indicate that no more than 40 subscribers at CO 1 want to talk with subscribers at CO 2 at the same time. Thus, only 40 lines are needed between the COs.

The process of deciding just how many lines are needed between locations is called *traffic engineering*. This has two main components: numbers and duration. During a 24-hour period, it might be possible that 400 subscribers want to talk with 400 other subscribers serviced by a different CO. However, if only 40 want to talk to others *at the same time*, only 40 lines are needed. As the duration of each call increases, the need for more lines also increases. If each of the 400 subscribers wanted to talk for 24 hours, then 400 lines would be needed.

This is the problem networks are presently facing. The infrastructure was designed based on a certain number of subscribers with a certain average call duration. The number of subscribers has increased primarily because almost everyone now has telephone access, but also because of the large increase of lines per person with the use of fax lines, "second lines," and dedicated lines for other purposes) but, more importantly, the duration continues to increase. New communication technologies which make use of the existing infrastructure cause problems for the operating companies in providing the same levels of service. This can cause "brown-outs" because there are not enough connecting lines to handle the demand for calls.

We are now faced with a situation where the existing infrastructure is insufficient to provide continuously increasing service at the new traffic levels. The long-range solution to the situation is to engineer new networks capable of supporting the increased traffic. The short-term solution, however, is to divert the new traffic (conforming to the new traffic duration needs) to a different network and eventually have that new network take over the duties of the old (or, perhaps, continue to exist in tandem but only for old services).

1.3.2 CIRCUIT-SWITCHES AND PACKET-SWITCHES

We said that a circuit is the connection which exists between two endpoints. However, it is only necessary to have the connection in place during the period in which it is in use. At other times, it would be preferable to use the connection for other purposes. This can be done only when the traffic is intermittent. Non-voice data transport falls into this category.

Data are often collected together into bunches called packets. The packet, like a piece of mail, has sufficient information within it to be distinguished from other pieces of mail. Also, like pieces of mail, it is possible for two (or more) pieces of mail to have the same address, and yet be from different senders. When packets have the same destination address (no matter the originator), they may be packet-switched. Say that two people want to send data to the same address. They must each have a separate line to the central switching office but, since they are both going to the same address, it is possible to use a single line to the destination. Three lines are used rather than four. This would not be possible if the data were continuous, but being packetized allows the line to be used for different end-to-end connections as long as the total amount of data does not exceed the capacity.

This also applies to subsets of the connection. For example: user A of Company B wants to send data to user Y of Company Z; user C of Company D wants to send data to user X of Company Z. It is possible (assuming the total data amount does not exceed capacity) for both packets to share the same line between the central office which services Companies B and D and that which services Company Z. However, at both ends, the packets must have their own lines to reach the final destination. Figure 1.6 shows that five connections are needed (to/from A, C, Y, X, and from CO B/D to CO Z) but one line is shared. This reduces the distance needed for the separate lines and reduces the infrastructure size (and, hopefully, the cost to the users).

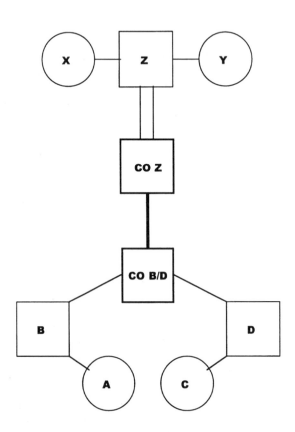

FIGURE 1.6   Central office line sharing.

An important point to notice in these shared connections is that it works only if the average data need is less than, or equal to, the capacity. "On average" is a term which requires some technical support. In cases where both users have data available at the same time, or when one (or both) user temporarily is using a larger amount of data than can be transported, something must be done to keep the present data load to the capacity of the line.

This is done primarily with buffers. Only one packet can be transmitted at a time. If two packets arrive at the same time, one must be stored until the other has been transmitted. Whenever the total amount of data arriving from the multiple endpoints exceeds the capacity of the connection, the number of buffers in use will continue to increase. If this never decreases, it is an indication that the network is under-engineered; the average data rate exceeds the capacity. However, if it is sufficiently well-engineered, the buffer pools will decrease once the total amount of data falls below the capacity.

We see now that a circuit-switched connection is dedicated between endpoints. A packet-switched connection can have parts of the connection shared between users wanting to transmit data between the same locations. The next subsection will discuss the degree of isolation between endpoints and the connection by the use of routers.

1.3.3 ROUTERS

A circuit is defined by the endpoints. A route is defined by the path that is taken between endpoints. Switching is the process of making a path available for use by a circuit. A router shifts data from one route to another.

In our general communications example, it would be possible to have a single line connecting all 1,000 subscribers. Use of such a line could be regarded as a "party line" where more than one subscriber is capable of using the line at the same time. However, if the data has been packetized, it is then possible for each subscriber to put data onto the line—just not at the exact same time.

1.3.3.1 LANs and WANs

Such a situation is known as a Local Area Network (LAN). While it is more likely to be found within a corporate environment, it may also be encountered in residential use where more than one device wants to access common resources. For example, two computers both want to share a printer. If both computers and the printer are on the same LAN, then the printer can be accessed by both computers (or the computers could share file systems located on their local storage) by making use of the LAN and packetized data.

Routers become useful when multiple networks are in effect. Routes may be permanent or temporary. A LAN which is always operational provides a permanent route. A route which may be set up when needed and torn down when no longer needed is temporary. A switched (circuit or packet) connection is a temporary route on a Wide Area Network (WAN).

The difference between LANs and WANs is primarily one of distance, but it is also one of topology. A WAN has varying routes depending on present network circumstances and needs. For example, a user in Denver needs a connection to Buenos Aires. At one time, the connection might be from Denver to Dallas to Mexico City to Buenos Aires. Another time, the connection might go from Denver to New York then by satellite directly to Buenos Aires.

The LAN, therefore, is usually a permanent, fixed route while the WAN provides a varying set of routes based on present needs and availability of resources.

1.3.3.2 Functions of the Router

A router must have address information associated with each packet. One of two general situations must occur; either each packet contains full origination and destination information *or* a special identification is set up for a particular origination/destination set on a temporary basis. The router will have "address tables" or a routing directory, which enables it to determine the path needed for the data. If User A wants to communicate with User B and they are both on the same LAN, the router does nothing (except to examine the packet). If User A wants to communicate with User F and they are on different LANs but the router has a direct connection (called a *node*) on both LANs, then the LAN has the duty of grabbing a copy of the packet from the first LAN and putting it onto the second LAN. Note that the data still exists on the first LAN but should be ignored by all nodes which do not have the destination address.

Routers are deemed particularly useful when they have access to WANs. User A wants to communicate with User Q. User A is on LAN 1. User Q is not even on a LAN. A router on LAN 1 can make a connection, through a WAN, to User Q (or vice versa) and provide a temporary access route. Figure 1.7 shows a variety of possible access routes.

To summarize, routers allow access to various fixed, or temporary, routes. They do this by recognizing how to get to specific destination addresses and copying data from one route to another. This is particularly useful in Internet applications and is also very useful when data of varying amounts must make use of limited resources.

1.4 MULTIPLEXING

Multiplexing is the process of putting more than one stream of information on a physical circuit at the same time. The two primary methods of doing this in transmissions are Frequency Division Multiplexing (FDM) and Time Division Multiplexing (TDM) (see Figure 1.8). The earlier radio example is a good one of FDM. Within a certain range, one broadcaster may transmit at a frequency of 500 KHz (+/-3 KHz probably). Another broadcasts at 510 KHz. Both signals can take place over the same medium (air waves) because there is no overlap.

The packet-switched network above is a good example of TDM. In this situation, a packet meant for one recipient is followed by another meant for someone else. As we will see in discussion on the various "flavors" of xDSL in the next chapter, this can also be more tightly delineated.

FDM and TDM can be used separately or in combination. Frequency multiplexing requires "guard bands" allowing for imprecise (or mildly distorted) transmissions.

Analog and Digital Communication

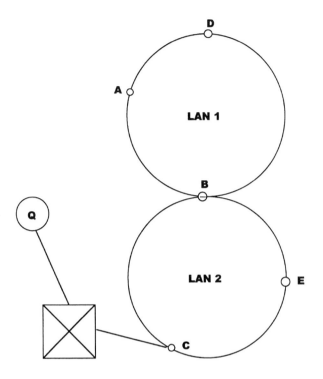

FIGURE 1.7 LAN and WAN routing.

TDM is a more precisely defined algorithm: defined at the micro or macro levels. At the micro level, each bit (determined by the sample taken at the defined clock rate) is routed to a specific physical or logical destination. At the macro level, the contents of the packet can be examined and routed according to the information content.

Multiplexing is also used to a great extent for long-distance lines ("trunks"). FDM works very well for separating circuits over the same physical medium and TDM contributes when packets are being routed over the line. The amount of multiplexing is used to determine the capacity and category of the long-distance trunks.

1.5 INFRASTRUCTURE LIMITS

Every infrastructure—regardless of whether it is a highway system, a telephone system, a power system, or something else—has a set of design criteria. These criteria say what is needed. The system may be designed to *exceed* the stated criteria, but there will still be limits.

In the case of the Public Switched Telephone Network (PSTN), these criteria were devised in accordance with the needs of using copper wiring to transmit speech. The first criterion is to have the necessary bandwidth. Human speech and hearing is able to utilize information of about 20 Hz to 20,000 Hz (a Hertz [Hz] is a unit indicating a cycle per second, in this case applied to sound waves). However, it is

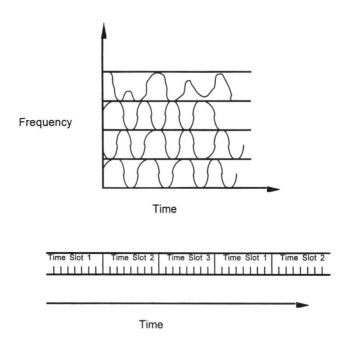

FIGURE 1.8 Frequency and time division multiplexing.

not necessary to use this entire spectrum for speech. Except for a few rare individuals, most human voices cannot generate speech signals greater than about 3,700 Hz (note that the criteria might have been a bit different if they were designed specifically for music). There *is* a difference between what is *possible* and what is *necessary*.

Thus, the wiring from the residential or business user to the central offices has been designed around a *passband* of 300 Hz to about 3.3 KHz (3 KHz range). For general use, it can be defined as between 0 Hz and 3.7 KHz. This allows a guard band above it for FDM purposes and keeps each channel to 4 KHz.

This 4-KHz number is very important in other design criteria for later, digital, uses of the network. As we saw above, Nyquest's Sampling Theorem says that we must sample at twice the rate of the signal change (or information rate). This means that 8,000 samples per second must be taken to digitally sample voice. If we say that we can define the sample's value with 8 bits of data, then a voice signal can be represented digitally in a data stream of 64,000 bits per second.

Since the transmission infrastructure was designed with speech in mind, the primary criterion was to carry speech signals clearly. This meant that various methods were devised to ensure that the problems of degradation and attenuation were addressed. This meant that loading coils and repeaters were designed with the 3.3 KHz spectrum in mind, as described in the next subsections. Also, to allow full use of the lines, bridged taps and Digital Loop Carriers (DLCs) were implemented for use on the local line (or "loop").

Analog and Digital Communication 15

1.5.1 Distance Limitations on Local Loops

The local loop, as discussed in Section 1.2.1, is limited by problems with degradation and attenuation. This is caused by a combination of factors: thickness, impurities, bridged taps, etc. The electrical factors which come into play most often are called *resistance, inductance, capacitance,* and *admittance*. In electrical formulas, the symbols used for these factors are R, L, C, and G, and the factors are therefore referred to as RLCG parameters.

In non-mathematical terms (therefore not precise) capacitance is the ability to store electricity within the material. Capacitance tends to vary most depending on the material (gold less than copper, for example). Resistance is the "stickiness" of the material, the tendency to prevent the electricity from flowing through the material. Resistance tends to vary depending on the frequency (with greater resistance at higher frequencies). Inductance is concerned with the tendency of the material to convert the energy into magnetic fields and is a combination of material composition and thickness. Admittance is the reciprocal of impedance and is the ratio of voltage to current.

These factors are important in determining transmission characteristics for electrical lines. They are also important in determining just what needs to be done to the line in order to improve performance for specific needs. The precise methods of use, however, are beyond the scope of this book.

The important aspects pertaining to this discussion are that the wire gauge, purity of material, frequencies transmitted, and manipulations of the physical medium (twisting, shielding, devices to shift characteristics) act to limit the distances various lines can be used. Distances supported range from 1,000 feet for VDSL to 18,000 feet for most lines running at 128 kbps or less. As will be discussed in the next subsections, loading coils and repeaters can extend this range, but do so by shifting the transmission capability toward the voice spectrum (thereby causing problems with xDSL techniques seeking to use spectrums above 4 KHz).

1.5.2 Loading Coils

Loading coils are useful in reducing attenuation because capacitance and inductance interact with each other causing shifts in the voltage and current phases for the passage of electrical signals. The capacitance cannot readily be changed, however the inductance can be *increased* to better synchronize the current and voltage phases (thereby reducing power requirements and decreasing attenuation).

The devices used for this purpose are called *loading coils* (usually iron rings). When the loading coils are wrapped with the Unshielded Twisted Pair (UTP), there is an increase in the inductance in the line. The effect varies according to the spacing and number of the wraps. They are designed with specific frequency bands in mind and although they increase the potential transmission distance for the spectrum for which they are designed (normally speech), they decrease the capability to transmit over other spectrums.

1.5.3 Repeaters, Amplifiers, and Line Extenders

Degradation and attenuation can be improved by changing the physical characteristics of the transmission line. They can also be changed by treating the extended line as if it was a series of shorter lines. In the analog world, this is done by using *line extenders*. A line extender is basically an amplifier (similar to a loudspeaker used for a human voice). It takes the signal and adds power to it, effectively resetting the distance marker back to 0.

However, since analog signals are continuous, there is no effective way to recognize errors in the signal and it is impossible to eliminate them once the signal has been carried further along. An extended line can only cause the length to be increased, but any errors that enter into the signal will continue to be compounded. Loud music can be heard at a much greater distance than soft music, but the clarity and ability to be understood will continue to degrade with distance.

A repeater is a digital, or hybrid digital/analog, device which attempts to recreate the signal. It can only be designed for known signal patterns since it must have the "knowledge" of what parts of the signal are likely to be in error. Repeaters are generally used with DLCs (discussed later). Since it is designed in terms of specific requirements, it must often be replaced when the line is to be used for new protocols and line extenders must be removed or replaced with appropriate repeaters.

1.5.4 Bridged Taps

We have been discussing local loops as if they were a single uninterrupted line extending from the residence or business directly to the central office. While this is sometimes true, it is also quite possible for there to be a number of places where there is a 'T' junction in the line. This may be because the line was formerly (or currently) used as a party line, or it could be a result of extensions of the line within the environment.

Similar to matching mixed wire gauges, a bridged tap can cause echoes and other transmission difficulties. With modern electronic solutions and adaptive physical protocols, this can usually be solved for xDSL, but it is a concern for equipment design as well as standardization.

1.5.5 Digital Loop Carriers (DLCs)

Use of Digital Loop Carriers (DLCs) addresses two problems within the area served by the central office: the first is distance and the second is better use of limited facilities.

Almost all long-distance trunk activity is digital in nature. As implied in earlier discussions, it is much easier to carry digital signals for long distances without severe degradation of signal quality than it is to transport analog signals. A DLC takes this digital link and brings it closer to the residence or business subscriber. This reduces the analog distance requirement allowing for better rural service.

Currently, a set of four wires (two twisted pairs) can be used for larger capacity digital transmission (discussed more fully in Chapter 2). Thus, DLC can also be used to reduce the total number of twisted pairs required for additional line service.

Analog and Digital Communication 17

Two twisted pairs can service 24 lines, giving a net reduction of 22 twisted pairs for the area that the DLC is used.

Whatever the reason or benefit for using DLCs, it creates a problem when the line is to be used for a specific subscriber xDSL. The DLC expects analog and will be converting into a specific digital signal. This is a problem. Estimates run as high as 30% use of DLCs on the present network.

1.5.6 Summary

Analog transmission makes use of loading coils and line extenders, or amplifiers, designed for specific frequency use. Analog or digital lines may have bridged taps into the line which cause extra complications for signaling protocols. Finally, use of DCLs can prevent use of arbitrary xDSL protocols on subscriber lines. All of these factors can preclude use of xDSL protocols on the line, but are very useful for analog transmission purposes.

1.6 BOTTLENECKS

A bottleneck is a reduction in size causing a limitation on flow. In fluid mechanics, a reduction in size with a constant pressure may cause the speed of the contents within a pipe to increase. That is, if one unit value per second is being forced through a pipe, the reduction in size of the pipe causes the distance per unit time to increase to accommodate the net volume. This doesn't work in the area of data transmission. The speed of transfer will be limited to the slowest part of the connection.

There are a number of areas that affect transmission capabilities. These can be further broken down into subareas, but the major areas are sufficient for purposes of this discussion. These areas include: host I/O capacity, access line capacity, long-distance capacity, network saturation, and server (or far-end peer) access line and performance. Figure 1.9 shows the components of the end-to-end transmission line.

1.6.1 Host I/O Capacity

Host I/O capacity is the ability of the host (local computer) to transfer and manipulate data. This capacity is affected by several components including processor clock speed and instruction processing ability, amount of RAM available, disk I/O transfer times, data bus or I/O port capability, and xDSL access device efficiency. A slow processor may be unable to utilize large data transfer rates. RAM limitations may prevent applications (such as a browser) from operating to speed. Disk I/O transfer time is important when files are being transferred from one file system to another. A data bus, or I/O port, has a specific designed transfer rate. The old RS-232C "serial ports" are often limited to no more than 56 kilobits per second (kbps). Newer serial ports may have transfer rates greater than 300 kbps (still slow in comparison to potential xDSL transfer rates). Finally, the device (often referred to as a "digital MODEM") which provides xDSL access to the subscriber line must be capable of handling the data rates.

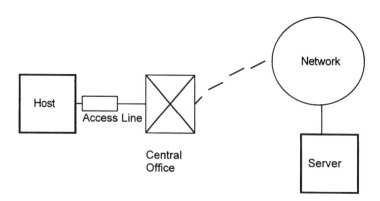

FIGURE 1.9 Potential bottleneck locations.

What this means is that in order to make use of faster data transfer technologies, an upgrade may be necessary. If you are using a 780 kbps access technology and your computer can only handle 128 kbps, you will not see any improvement over that provided by a 128 kbps access technology.

1.6.2 Access Line Capacity

Access line capacity is the primary focus of this book. The various Digital Subscriber Line (xDSL) protocols give a specific range of *possibilities* for access speeds. The condition of the line between the subscriber and the central office (or central data server location) will affect capacity in one of two ways. It will either change the error rate—the need to retransmit data corrupted by line conditions—or it will reduce the amount of data that can be transmitted. Many of the xDSL technologies have rate adaptive variants to allow the use of the line to be changed according to current line conditions.

The choice of access technology will affect total transfer capability. Use of analog limits transfer to 56 kbps of uncompressed data at present (but will normally be less). Digital protocols will allow greater speeds depending on the line conditions; e.g., cable modems make use of shared coaxial to multiplex access to Internet Service Providers (ISPs). If there is one user on this shared medium, the speed may be very fast. If many users are sharing the medium, speed may be less than that available with analog technologies. xDSL Technologies which map directly into speech circuits are less likely to be influenced by line conditions (because the infrastructure was designed for speech). Others, however, are directly affected.

1.6.3 Long-distance Line Capacity

In most cases, the subscriber will be connecting to an endpoint that is *not* serviced by the same central office. This means that long-distance trunks will be in use. Some of the design aspects of ADSL and other xDSL technologies take this into account by incorporating high-speed signaling protocols within the stack used over ADSL.

Analog and Digital Communication   19

Discussion of the DSL Access Module (DSLAM) will focus on different methods to gain access to sufficiently high-speed networks, allowing for a more productive use of the access line speed. At any rate, the long-distance network must have at least the same transfer capacity to not limit the throughput.

1.6.4 NETWORK SATURATION

If we are entering into a packet-switched (or routed) network, the total amount of traffic will determine the amount of data capacity available to each subscriber. The World-Wide Web can easily turn into the "World-Wide Wait" when many users are trying to make use of the network at the same time.

1.6.5 SERVER ACCESS LINE AND PERFORMANCE

Performance factors involved with the server are very similar to that of the host, with one addition. The server is likely to be providing access to multiple subscribers at the same time. Thus, the performance of the server, combined with number of access lines and the number of presently active subscribers, determines the capacity of the server.

1.6.6 SUMMARY

At present, the Internet access network is primarily limited by network congestion and server capacity. This limit is often less than the access line capacity. (It is variable, depending on the number of active connections and the specific server being accessed.) xDSL Technologies primarily provide a vehicle for future capacity. There will *always* be a bottleneck in a data flow system. The throughput is increased by improving each segment so that the capability of the *lowest* is increased.

In Chapter 2, we will introduce the various current flavors of xDSL technology. It will also give a certain amount of history of both technologies in use and the standardization needed to allow use of new protocols, such as ADSL, within an international network.

2 The xDSL Family of Protocols

The Digital Subscriber Line (DSL), as discussed in the introduction, utilizes digital information as the primary data form over the lines from the residential or business user to the central office (or central-line endpoint). The ADSL Forum indicates that the terminology, originally defined by Bellcore, is meant to apply only to the devices used on the line and, thus, DSL is meant to refer to a particular DSL device (often BRI ISDN, the first UTP technology applied to DSL use).

Whether this is historically true, DSL can be used to describe the line. The protocol, however, will be described by applying an adjective to the line type. Thus, Asymmetric Digital Subscriber Line (ADSL) and High-speed Digital Subscriber Line (HDSL) are part of the xDSL family. Some DSL technologies do *not* have DSL as part of their name. Examples of this are high-speed "analog" Modulator-Demodulators (MODEMs, often just referred to as modems) and earlier DSL technologies such as Basic Rate Interface Integrated Services Digital Network (BRI ISDN). A physical line protocol which provides digital data transmission to and from the residence or business local connection is part of the xDSL family.

2.1 FROM DIGITAL TO ANALOG

In the previous chapter we discussed the differences between analog and digital data communication. In the form of electrical signals, the two are closely related, with digital being a subset of analog. It is an important difference, however, as the discrete sampling techniques used for digital data transmission enable the use of methods which allow for long-distance, relatively error-free communication. The same is true of modems.

The first modems were a (relatively) simple translation of digital information to analog form (and back) so that the same 3-KHz speech bandwidth could be used for the transmission of data. The techniques and available hardware were rather "loose," allowing a greater amount of bandwidth than was theoretically needed for the data transmission speed. As modems improved, the mechanism shifted from translation to use of protocols. The Link Access Protocol for MODEMs (LAPM) provided a framing technique that allowed for better error detection and correction. At present, a combination of digital and analog techniques are used for "56K MODEMs." In actuality, it is rare for 56K modems to be able to provide full bandwidth.

Most subscriber lines are used for bidirectional information—data sent to from one end and received on the other end. Because of interference in the wiring within a home or business, as well as a few other physical factors, reception is usually better than transmission (this is also acknowledged in the setup of ADSL as we shall see later). A 56K modem is designed to allow for this, as well as the fact that almost all long-distance trunks make use of digital transmission. Received data that has been sent digitally to the place where the local loop starts (or even the end of the DLC) can take advantage of the digital transmission capabilities and the 56K modem can then translate this data with relatively few errors.

If an analog-to-digital conversion takes place over the line before the local loop, 56K communication is not possible. In the direction of the subscriber to the network, 56K communication is almost never possible. However, this is the best (due to limitations arising from the repeatedly mentioned Nyquest's theorem) that can be done on the 3-KHz voice band.

We see that, in several ways, analog merges into digital. Use of specific digital techniques as well as reliance on digital long-distance trunks provides a merging of coding schemes and can place high-end "analog" modems into the DSL class of protocol families.

2.2 DIGITAL MODEMS

The above 56K modems are still primarily analog. Digital "modems" do not actually provide modulation/demodulation of the digital information into analog wave forms. They preserve the digital form. They are primarily called modems to make the technology more comfortable to people who are accustomed to MODEMs. From a "purist" point of view, however, they are not modems. From another "purist" point of view, neither are 56K modems. The transition is a gradual one.

A digital modem can also be called a Terminal Adaptor (TA). A TA acts as an interface between a host computer system and the access line. There are two categories of TAs. The first is the "traditional" TA. This enables older types of equipment to make use of faster, newer, access methods. Usually, this is limited to serial ports (just like the analog modem is attached to the computer). A number of different protocols are available to transfer the asynchronous data through the serial port across the digital access line. The precise protocol is not important (except for speed limitations associated with the protocol) but, as is true for all other communication methods, it is important that both endpoints expect to make use of the same protocol.

The second type is more often called Terminal Equipment (TE). Although it serves the same function as a modem—providing host connectivity to servers—it is *not* really doing an adaptive protocol to allow old-style (analog expectant) terminal equipment to a new access line. TEs will often make use of direct-host connections by either using the system data bus or by a high-speed LAN interface.

2.3 THE ITU-T, ADSL, AND ISDN

Some descriptions of ADSL in the media allude to "ADSL is the replacement technology for ISDN." Statements such as this are only partially true. The ISDN to which these articles refer are for the lower speed Basic Rate Interface access method for Integrated Services Digital Networks (BRI ISDN). As we will see later in the book, BRI ISDN does provide a lower speed connectivity (at least, in the network to user direction) over the DSL than does ADSL. However, the use of ISDN to refer only to BRI is misleading.

Some books and media articles have reported that ADSL and ISDN are two different things. At the same time, these books and articles will indicate that Asynchronous Transfer Mode (ATM, known also as "Broadband ISDN") is a recommended part of how to use ADSL. As we will see later, ADSL was also designed to allow the possibility of carrying of BRI on top of ADSL. It can be very confusing to a technical user—as well as to the technical manufacturer! The fact is, (as listed in the ADSL forum for xDSL technologies and other places) that ADSL fits into the ISDN architecture, *but* with a significant difference which has primarily come about from the evolution of the Internet and the use of routers and the TCP/IP protocol. This difference will be discussed in this section, as well as the reasons why ADSL can reasonably be considered part of ISDN.

Long-distance trunk lines are almost always digital, and have been for quite a while. However, there have been upgrades to the network in recent years to increase data rates from 56 kbps to 64 kbps. ISDN evolved as an architecture to extend the digital network all the way to the home or business. The basic idea was that, for the highest data throughput, keeping the data in digital form from one endpoint to the other was very important. It was also felt that keeping "backwards-compatibility" was a requirement for such an upgrade to the international transmission infrastructure.

The International Telecommunications Union Telecommunication Standardization Sector (ITU-T, formerly referred to as the International Telegraph and Telephony Consultative Committee or CCITT) is a standing committee of the United Nations. The role of the ITU-T, in addition to other North American and European standards bodies, with respect to ADSL will be discussed in the next section. At this point, we will discuss the part ITU-T played in putting together the architecture for ISDN.

Figure 2.1 is from the ITU-T's recommendation I.325. As "the basic architectural model of an ISDN" it allows for various methods of access—both newer and older methods. Note that the figures are divided into three categories: 64-kbps capabilities, greater than 64 kbps capabilities, and signaling and specialized switching capabilities. Depending on its specific configuration, ADSL can be considered to be either ">64 kbps nonswitched," ">64 kbps switched," or ">64 kbps nonswitched" in conjunction with "packet switching," "common channel signaling" or "user-to-user signaling."

The difference between the use of ADSL and the ISDN is not one of exclusion—rather it is one of non-specific inclusion. As can be seen from Figure 2.1, nothing in ADSL is excluded from ISDN, however, the use of routers is not indicated as part of the architecture. "Signaling" indicates that a connection is to be set up. A router already has the transmission path (or paths) in place and uses the address information included as part of the packet to route it to the proper destination. Thus, ADSL can be seen as ">64 kbps routed capabilities." As we will see later, ADSL can, however, certainly *be* switched and contain signaling information, for a part of its transmission path.

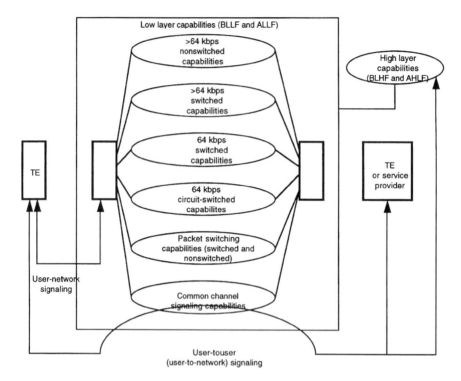

FIGURE 2.1 ISDN Basic architectural model. (From ITU-T Recommendation I.325.)

ADSL can therefore be considered to be a part of the ISDN architecture and its use certainly can incorporate other ISDN data and signaling protocols. Unlike the general ISDN architecture, however, it does not *have* to contain signaling information from the TE to the network. The dashed line from the TE to the service provider can be interpreted as the type of link provided from an ADSL unit to the DSLAM.

2.4 ADSL STANDARDIZATION

The physical layer, and higher protocols, must be coordinated between endpoints. However, this can be standardized within the company's manufacturing equipment or in an organization providing access services. Physical layer specifications for ADSL will be examined in Chapter 3. There is no requirement for a standard other than for matching within a transmission line's endpoints. A standard is useful so that companies can manufacture and use equipment that work with equipment of other companies. This solves the problems of localization of service (ie., you buy

The xDSL Family of Protocols   25

equipment that works in one location but, for very similar service, you find you have to buy different equipment if you move to a different locale). It also allows consistent access to public networks and services.

2.4.1 STANDARDS BODIES

There are four groups that contribute to standards. The first is at the corporate level, where a product is developed in accordance to perceived market needs. This standard can be proprietary, so that only the company devising the standard can make equipment based on the standard. It can also be *open*, which means that anyone who wants to make use of the standard is able to do so. Sometimes, the standard is so widely accepted by the public that the company's standard becomes an industry standard. This is called a "de facto" standard. The 'AT' command set formulated by Hayes is one such standard.

Sometimes, the perceived market is large enough that companies, universities, and other interested parties will group together. This allows them to use the abilities and experience of all of the members to create a standard that is more flexible and, with the endorsement of the entities involved, more likely to make an impact on the general industry. The ADSL Forum and ATM Forum are examples of this category.

The next level of committee is at the national, or regional, level. The American National Standards Institute (ANSI) is one such body in the United States. The European Technical Standards Institute (ETSI) is a similar body in Europe.

The final level is that of the ITU (or, for things other than telecommunications, some other body of the United Nations). The ITU does not have the authority to mandate standards to the various countries and companies that may eventually make use of the protocols. Rather, they issue recommendations and these recommendations are adapted by the various national, and regional, committees which do have the ability to make sure that products meet their standards. In unregulated markets, such as the United States, the final version is that adopted by the specific manufacturer.

2.4.2 ADSL STANDARDS BODIES

Four groups are the most active in setting up ADSL standards. At the first level, semiconductor and other device manufacturers have been involved in worldwide test trials. The most active manufacturers have joined together as members of the ADSL forum (http://www.adsl.com). Some of them have also created a subgroup of the ADSL forum called the Universal ADSL Working Group (UAWG) which is concerned with simplifying ADSL to make it a more consumer-friendly product.

The next groups are at the national level. They are ANSI and ETSI. ANSI has published the primary ADSL documents to date, with input from ETSI. Finally, the ITU has become involved to coordinate international use of ADSL so that it can be properly integrated into global infrastructure changes. Table 2.1 gives a list of standards and the standardization bodies involved. The following sections describe the contributions made by each group.

TABLE 2.1
ADSL Standards Bodies and Standards

Standards Body	Working Group	Standard	Purpose
ADSL Forum	Many, including UAWG SNAG	TR-00x	Industry Advisory Papers
ANSI T1	T1E1.4	T1.413, Issue 2	Basic ADSL Standard
ETSI	TM6		Close interworking with ANSI
ITU-T	Study Group 15	G.992.1	International Standards
		G.992.2	
		G.994.1	
		G.995.1	
		G.996.1	
		G.997.1	

2.4.2.1 ADSL Forum and UAWG

The ADSL Forum (http://www.adsl.com/), formed in 1994, has been in the forefront in publicizing ADSL and acting as a working group to explore architectural issues not yet covered by the standards bodies. It also acts as a "prod" to the standards bodies to get them working on standards. Most of the technical issues, addressed by the industry participants of the forum, deal with interworking. There are four subgroups that are particularly active. These are the UAWG (mentioned above), the System Network Architecture Group (SNAG), the Management Information Base (MIB) group, and the test group which coordinates test environments between forum participants and other manufacturers of equipment. Many of the issues being looked at by the ADSL Forum are also of interest to the Asynchronous Transfer Mode (ATM) Forum (http://www.atmforum.com/).

2.4.2.2 ANSI

The American National Standards Institute (http://web.ansi.org) oversees various committees composed primarily of industry technical people. The committee T1 and, more specifically, the subcommittee T1E1 is associated with ADSL standards. The T1E1 subcommittee has the responsibility of overseeing standards work for interfaces, power, and protection for networks.

Each subcommittee may be broken down into working groups. The T1E1.4 working group is actually the group responsible for DSL access, including ADSL. DSL access includes physical layer standards and transmission techniques for interfaces. The T1E1.4 working group was responsible for the T1.413 ADSL standard, which will be covered in detail in the next chapter. T1.413 was published in 1995. A new version (T1.413, release 2) is now available, in draft form, which incorporates some of the UAWG simplification issues.

2.4.2.3 ETSI

ETSI (http://www.etsi.fr) is also broken down into groups and subgroups. The Transmission and Multiplexing (TM) group contains the working subgroup TM6 which roughly corresponds to T1E1.4. TM6 often works with T1E1.4, making sure that international issues are addressed.

2.4.2.4 TU-T

The International Telecommunications Union (ITU, http://www.itu.int) is sometimes the last body to get involved with a new technology. However, the fact that the ITU-T did join ADSL standardization efforts in 1998 is an indication of the growing international desire to incorporate such services. The ITU-T generates technical recommendations, such as I.325 mentioned above. The present set of ADSL-related recommendations being worked on by the ITU-T are named by 'G' prefixes, as they fall into physical layer protocol categories. Recommendations currently under study are G.DMT which is largely a rewrite of T1.413, G.lite which incorporates much of the work of the UAWG, G.test which concerns test specifications for xDSL, G.OAM concerning operations, administration, and maintenance aspects of xDSL and G.HS for handshaking protocols to allow startup negotiation.

G.lite has now been informally accepted by the ITU-T and will be voted on during the next meeting of ITU-T Study Group 15 to be held in Geneva, Switzerland in June 1999. It is currently available via special user's groups, such as the ADSL Forum. The standard is not expected to change much in that process although nothing is guaranteed. The versions of G.DMT, G.OAM, G.HS, and G.lite will be released as G.992.1, G.992.2, G.994.1, G.995.1, G.996.1, and G.997.1. Specifically, the ITU-T recommendations at present are:

G.992.1 (G.dmt)	Asymmetrical Digital Subscriber Line (ADSL) Transceivers
G.992.2 (G.lite)	Splitterless Asymmetrical Digital Subscriber Line (ADSL) Transceivers
G.994.1 (G.hs)	Handshake procedures for Digital Subscriber Line (DSL) Transceivers
G.995.1 (G.ref)	Overview of Digital Subscriber Line (DSL) Recommendations
G.996.1 (G.test)	Test procedures for Digital Subscriber Line (DSL) Transceivers
G.997.1 (G.ploam)	Physical layer management for Digital Subscriber Line (DSL) Transceivers

2.5 THE xDSL FAMILY OF PROTOCOLS

The ADSL Forum refers to the protocols covered in this section as "Copper Access Technologies." Although pure analog modems are also included in this category, we will stay with the primarily digital access devices, starting with 56K modems to Very-high data rate DSL (VDSL).

Table 2.2 shows the various protocols and their important attributes. Most media articles focus on data rate, but signaling and infrastructure are also very important to the consumer and the manufacturer. This is because these give an idea of the flexibility of the physical interface. The "best" interface will vary depending on the needs of the application (as well as the cost structure associated with the service).

This is a good time to discuss one of the most frustrating and uncertain aspects of all of the xDSL technologies: cost. Use of 56K MODEMs has the greatest advantage because it uses the same infrastructure, and rate mechanisms, as does speech. In North America, and other regulated locations, speech service is usually kept at a low cost level to help provide "universal" access to the service.

However, anything other than "regular" speech lines is an opportunity to ask for special usage rates from local, or national, regulatory committees for telecommunications network providers. On the one hand, telecommunications network providers (call them RBOCs as a shorthand, although this is not really applicable outside of North America) have a great need to be able to expand their infrastructure and profitably be able to support a network with a rapidly skewed traffic specifications. Thus, they will request high rates for any xDSL which makes use of existing infrastructure (BRI ISDN, PRI ISDN, some SDSL). These rates are able to subsidize losses from overutilized (ie., not within the traffic engineering duration guidelines) analog speech lines. Unfortunately, such rates do not provide a good cost basis for use of the xDSL technology. If a BRI ISDN provides three times the bandwidth of a POTS analog circuit, then the user will not want to pay *more* (preferably less) than three times the cost of a POTS analog circuit. Unfortunately, this is not the case—BRI ISDN may cost much more than that multiple in one state and barely more than a single analog line in another state.

So, here comes the other category of xDSL protocols—those that don't make use of the existing speech network infrastructure. The RBOCs can argue for an access cost which is *less than* the cost of an analog line because it does not cause any loading on the existing infrastructure. Any fee basically becomes a rental fee on the local loop. The RBOC may also provide the data service so that the total cost may be a multiple of analog service costs—or the "bare copper" may be leased by an Internet Service Provider (ISP) and the user recharged for the local loop line "rental" as part of their Internet access fee. Both of these techniques allows the RBOC to keep the load off of the infrastructure *and* to make a higher profit on the xDSL service than can often be negotiated with regulatory bodies on traffic that is carried on the PSTN.

Thus, the final answer on cost versus value for the various xDSL is a big question mark. Many of the user newsgroups are indicating that price is the biggest component in residential decisions—more than value represented by price versus speed. The decision to use ADSL, or one of the other xDSL protocols, will depend on the specific service and global connectivity needs plus the costs of the various services in the consumer's specific area.

TABLE 2.2
Digital Subscriber Line Technologies

Name	Meaning	Data Rates	Signaling	Infrastructure	Comments
V.90	Int'l 56K MODEM	56K downstream 33.6 upstream	Analog	Speech network	Analog/digital hybrid
BRI ISDN (DSL)	Basic Rate Interface for ISDN	Up to 128 kbps + 16 kbps X.25 (160 kbps raw)	Digital Q.921/Q.931	Speech network	Extension of digital long-distance network to customer premises
IDSL	ISDN Digital Subscriber Link	128 kbps	None (BRI ISDN spoofed)	Direct links to ISPs or IP routing	Avoids use of speech network but uses standard BRI ISDN equipment
HDSL/HDSL2 (SHDSL) PRI ISDN	High-data rate Digital Subscriber Line or Primary Rate Interface for ISDN	1. 544 Mbps (T1) 2. 048 Mbps (E1)	Channel-Associated Signaling or Digital Q.921/Q.931	Speech network	HDSL requires two UTP (T1) or three UTP (E1). HDSL2 uses only one or two UTP
SDSL	Single-line Digital Subscriber Line	774 kbps	Channel-Associated Signaling	Speech network	Uses HDSL technology with only one UTP Sometimes HDSL2 (SHDSL) is called SDS
ADSL/RADSL	Asymmetric Digital Suscriber Line Rate-adaptive Asymmetric Digital Subscriber Line	1.5 to 6 Mbps downstream 32 to 640 kbps upstream	None or ATM (Q.2931)	Direct links to Service Providers or Cell; packet or frame switched or IP routing	Most ADSLs are rate-adaptive Service highly dependent on line conditions

Name	Meaning	Data Rates	Signaling	Infrastructure	Comments
CDSL/ADSL "lite"	Consumer Digital Subscriber Line or reduced ADSL parameters	64 kbps to 1.555 Mbps downstream 32 to 512 kbps upstream	None or ATM (Q.2931)	ATM Cell Relay	Aimed at a "simplified" consumer-oriented MODEM replacement technology with reduced performance, reduced options, and easier and less expensive installation.
VDSL	Very high data rate Digital Subscriber Line	13 to 52 Mbps downstream 1.5 to 2.3 Mbps upstream	None or ATM (Q.2931)	SONET? (Optical ATM Cell Relay)	Distance limitations require that the local loop be connected into high-speed physical medium.

2.5.1 56K MODEMs

56K Modems were designed to squeeze the last of the bandwidth out of the 3 to 4 KHz bandwidth available from the speech. To do this, they had to make assumptions. The major assumption was that the long-distance network was all digital. This assumption meant that the only part of the loop that had to be optimized was the local loop from the Customer Premise Equipment (CPE) to the connection into the central office loop (whether at the central office or the end of a DLC).

The other major assumption, mentioned previously, is that the network-to-user direction (reception) was clearer than the transmission direction. This allowed for the possibility of 56,000 bps reception and 33.6 kbps transmission (according to ITU-T Recommendation V.34).

In order to achieve the 56,000 bps reception/download rate, a protocol was needed to map the Pulse Code Modulation (PCM) signal used over the digital network to a range of analog signals that could be carried over the local loop and be reconverted to digital at the user's equipment. Two major standards developed called K56Flex and x2. These standards are incompatible, and give a very good analogy to what has been happening in the ADSL market.

The first step in bringing a new technology to market is to make it work. This is usually done within a research environment. It may also be tested in various locations to give the technology a chance to be used in "real" situations. The second step is to bring the (usually proprietary) protocol or product to the general commercial market. The final step is for the product to reach either a "de facto" standard level *or* for the national, regional, and international standards bodies to determine that the technology is sufficiently important that a general open standard is needed. ITU-T Recommendation V.90 is presently the international standard for 56K supporting modems.

Note that 56K modems only address speed. They don't offload the switching system or improve traffic engineering situations. Also, the assumptions under which 56K modems have been set up can be misleading. Thus, a downstream (from the network to the user) speed of about 41 kbps is more typical for the user. In order to achieve more robust line speeds, it is necessary to bring the digital information stream all the way from one endpoint to the other.

2.5.2 BRI ISDN (DSL)

The first digital subscriber line technology devised for consumer access was that of Basic Rate Interface ISDN. As is true with the other xDSL technologies (even 56K Modems with their digital transmission line assumptions), the line must be capable of passing information on a bandwidth greater than that designed for the speech network. The process of making the local loop acceptable for more than speech is sometimes called "line conditioning." This process may include removal of loading coils, change of line-extender equipment, and removal of DLC segments. About 80% of the local loops in North America are expected to be able to be digitally usable.

The objective of BRI ISDN was to bring the digital part of the long-distance network, over a single twisted-copper pair of wires, all the way to the business or residence. A side benefit of bringing digital to the user was to have access to greater bandwidth (up to 144 kbps aggregate on BRI ISDN) for data transmission. Since both data (previously primarily MODEM coded data) and speech occupy the same physical media in the long-distance network, this meant that the user had access to both data and speech capabilities. These capabilities are referred to as "Bearer Capabilities" and the network has the opportunity to know the capabilities of each piece of equipment on the WAN and to allow, or disallow, connections between pieces of equipment based on tariff or compatibility issues.

This important point is worth restating: *BRI ISDN provides the same capabilities as current analog service with the addition of greater bandwidth access.* A BRI ISDN Terminal Adaptor can also provide one or two POTS ports, which allows use of the same phones, faxes, and modems that a user already has. The effect of the POTS port is to shift the analog-to-digital conversion from the central office to the customer's location

The only significant physical difference is the equipment used at the customer premise and the "line card" (a computer circuit board which supports one, or more, physical lines) in the switch. This is significant largely because of compatibility issues and cost issues. (Any change in the existing physical configuration requires capital outlay.) A person using BRI ISDN can place a call to anyone that they can presently connect to with analog service. (However, enhanced capabilities require both ends to be digital.)

This same service cannot be provided by ADSL as it does not connect into the same switching system. As we will see shortly, it may be possible to give the same type of global access using HDSL2 or SDSL with the same twisted pair.

This is the great advantage, and disadvantage, of BRI ISDN (and PRI ISDN, serviced by HDSL2 or SDSL). Using the same switching system provides equivalent global access, but also increases the burden on the current infrastructure. It will be possible to have the same type of access, and service, with ADSL by using a different switching system and parallel infrastructure, but probably not for many years. At present, only the technologies referred to as ISDN interfaces (BRI and PRI) allow for global connectivity to all existing, and new, telecommunications users.

2.5.2.1 Physical Layer

All of the ITU-T ISDN (including ADSL) refers to access points of the network. As seen in Figure 2.2, these access points are referred to by letters. The 'U' interface point is the place where the twisted pair enters into the business or residence. The 'R' interface point allows access of analog (non-ISDN) equipment to the line. The 'S' and 'T' interface points (often referred to as 'S'/'T') give access to the publicly defined physical interface. The 'U' interface, for BRI ISDN, can vary from one country, or region, to another (this allowed the RBOCs to provide the physical interface best for their existing networks). Therefore, we will concentrate on the 'S'/'T' interface.

In Chapter 1, we discussed analog and digital coding mechanisms. On the line leading from the central office to the 'U' interface point, a coding mechanism known as 2B1Q is used on the line. This gives quaternary ('Q') coding on the line and

The xDSL Family of Protocols 33

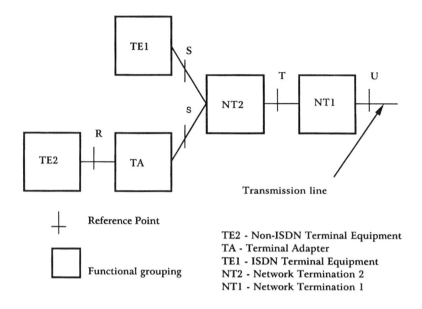

Note that xDSL makes use of equivalent reference points

FIGURE 2.2 ISDN reference points. (Adapted from ITU-T Recommendation I.411.)

helps to reduce the spectrum needed on the physical line. On the publicly defined interface (between the 'S'/'T' interface and the user's digital equipment), a method known as "pseudoternary" is used (where the null voltage level is interpreted as '1' and the high and low levels are '1' but are alternated to help balance the electrical characteristics. 2B1Q and pseudoternary coding are shown in Figure 2.3.

The next step is to organize the bits on the physical line into logical frames. For BRI ISDN, these are as shown in Figure 2.4. The use of a high-level '1' or a low-level '1', as defined within the pseudoternary coding mechanism, helps to determine the beginning and end of the frame. (Note that a frame and packet can often be used equally; a frame is really more of a physical encapsulation while a packet is a logical grouping within a physical environment.)

The full rated speed of the BRI ISDN between the 'S'/'T' interface and the user equipment is 192,000 bps. Note, however, that the data rate of the 'U' interface line is only 160,000 bps—allowing for the 144,000 bps of D- and B-channel data plus some simpler framing and maintenance bits. Since the 'U' interface line uses a quaternary coding method, however, the needed bandwidth is halved.

For the exact uses of the various bits on the BRI ISDN frame, the interested reader is referred to the non-ADSL ISDN books mentioned in the references. For our discussion, the only remaining point of interest in the physical layer is that of the TDM channels. These are referred to as the D-channel and B-channel 1 and B-channel 2. The 16,000 bps D-channel is used for signaling and for some data service (presently only X.25 is defined) provided directly by the network. The B-channels

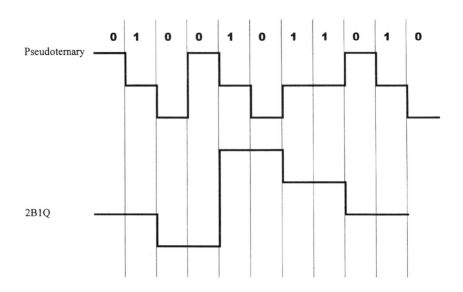

FIGURE 2.3 2B1Q and pseudoternary digital encoding.

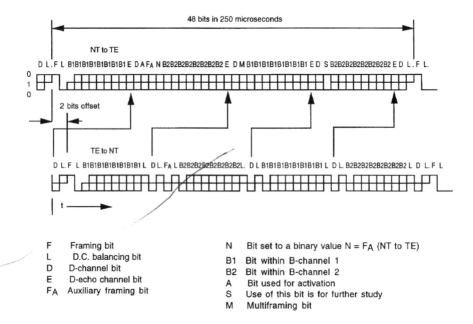

FIGURE 2.4 BRI ISDN framing from 'S/T' reference points. (Adapted from ITU-T Recommendation I.430.)

The xDSL Family of Protocols

(often referred to as B1 and B2) each provide a 64,000 bps data channel through the network (although whether it makes it all the way depends on the far-end's equipment).

2.5.2.2 Switching Protocol

The switching protocol used for BRI-ISDN is primarily defined by two ITU-T recommendations. These are Q.921 and Q.931. These define the protocols used for Open Systems Interconnection (OSI) levels two and three. The OSI model is one that is used for most modern protocols to foster *interworking* (access from one protocol, or network, to another). This is sometimes referred to as *internetworking* and, as you can easily guess, this is how the Internet was named. The OSI model will be discussed in more detail in Chapters 4 and 10.

The Q.921 protocol is referred to as the "Link Access Protocol for the D-channel" (LAPD). This is a High-level Data Link Control (HDLC) protocol and, as such, has five main components as seen in Figure 2.5. These are the "flags," an address field, a control field, an optional set of data carried by the frame, and a field which provides a limited degree of error detection and correction.

The "flag" is sent to mark the beginning (and end) of a frame. For HDLC, this pattern is '01111110' (hex 0x7E). Between the ending flag of one frame and the

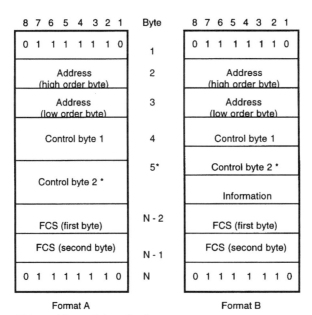

Format A
Format B

Format A is used for non-information frames
Format B is used for information frames (I, UI, FRMR, XID)

Control byte 2 is needed for frames with sequence information (I, RR, RNR, R. REJ)

FIGURE 2.5 Q.921 (LAPD) general frame format. (Adapted from ITU-T Recommendation Q.921.)

beginning flag of another flag, there may be either a fixed data value (continuous '1's are common) or a repetition of HDLC flags. The address field indicates the logical entity of the intended frame. The control field indicates frame type and, for some control types, indicates that a set of data may follow the control field. The final field before the "closing flag" is the Frame Check Sequence (FCS) which is a generic term for a numerical method that uses the remaining data to calculate a number to help verify integrity of the data. For LAPD, the FCS is done by a 2 byte Cyclic Redundancy Check (CRC).

Within the LAPD frame can be contained signaling information or specific protocol data. When it contains signaling information, it will be of the form specified in ITU-T Recommendation Q.931. The general message format is shown in Figure 2.6. The first byte, the protocol discriminator, allows mixtures of protocols within the D-channel. For Q.931, it is defined to have the value of 8. Using the value of 8 for the protocol discriminator defines the syntax of the rest of the contents of the frame.

The next byte will indicate the length of the Call Reference Value (CRV). The CRV will either uniquely identify a specific data or speech call (active or in the process of being connected) or will be "global" (specified by a zero-length CRV or a value of '0'). The CRV, if any, follows and then the message type is defined. The rest of the packet depends on the value of the message type.

The message types involved with Q.931 are of three categories. These are call setup, call teardown, and informational messages that may occur at any time. For example, the message types SETUP, SETUP_ACKNOWLEDGE, CALL_PROCEEDING, CONNECT, and CONNECT_ACKNOWLEDGE are the most important in call setup. DISCONNECT, RELEASE, and RELEASE_COMPLETE are the important messages during call teardown. INFORMATION and PROGRESS_INFORMATION are important during the lifetime of the call. Other details about call processing are important within the equipment and the interested reader is referred to ISDN-specific books such as those listed in the references.

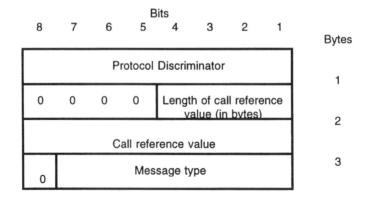

FIGURE 2.6 Q.931 message packet header. (From ITU-T Recommendation Q.931.)

2.5.2.3 Data Protocols

Signaling is useless unless some form of data is transmitted after the connection has been made (or, in some special cases, *while* the connection is being made). Support of various data protocols falls into two categories. We can call them "old" and "new." The "old" data protocols existed pre-ISDN and it may be necessary for the ISDN equipment to support them via the 'R' interface. One group of these protocols is the "modem protocols" which use an asynchronous series of characters with data and command modes. Old communication programs made use of de facto 'AT' command sets to control the modem and make sure that the modem was in the proper "phase"— data mode to transmit and receive data and command mode to change the characteristics of the modem or to perform call signaling.

Two of the popular asynchronous "adaptive" protocols, used with asynchronous character streams, are called V.110 and V.120. Both of these are ITU-T Recommendations. V.120 is used primarily in North America while V.110 is often used in Europe. A more popular protocol of late (due to the rapid ascension of the Internet) is that of MultiLink Point-to-Point (ML-PPP) protocol. This encapsulating protocol is very helpful in interworking and in helping to aggregate the bandwidth on multiple B-channels.

The faster frame-oriented protocols require either direct OS access or redirection to a LAN. This type of protocol includes X.25, Frame Relay, and ML-PPP (there are two main flavors of ML-PPP, asynchronous via a serial port and synchronous via a host OS application process). X.25 is an older protocol with many established global networks and is particularly used in Europe. In North America, Always On/Dynamic ISDN (AO/DI) has increased interest in X.25 because the networks may offer dedicated X.25 access on the D-channel. This allows a user to have a continuous connection to the Internet but does not occupy a B-channel (within the switched network) unless the additional bandwidth is needed.

Frame Relay and synchronous ML-PPP allow use of either broader "pipes" or aggregated B-channel data bandwidths. Many of the same techniques are useful with ADSL, depending on the exact configuration chosen.

2.5.3 IDSL

ISDN Digital Subscriber Line (IDSL) is a misnomer since BRI ISDN *is* a Digital Subscriber Line technology. However, since ISDN really applies to the digital network architecture, we can pretend that the term was created with the architecture in mind rather than the access method.

The idea behind IDSL is a simple one: make use of the slowly advancing ability of the Regional Bell Operating Companies (RBOCs, for North America) or PTTs to provide BRI ISDN service, but take the data load off the older infrastructure (and, in some areas, provide a cost relief to the consumer) by not requiring access to the PSTN at the central office.

The user is able to purchase a regular BRI ISDN piece of equipment. They then connect it to their ISDN line as if they were going to do a full BRI ISDN connection. However, at the central office, any switching messages on the D-channel are intercepted and responses, if needed, are "spoofed" such that the user equipment remains satisfied. The BRI ISDN equipment is thus able to make use of 128 kbps (both B-channels, either aggregated by the hardware or software combined by ML-PPP) or 144 kbps using both B-channels and the D-channels. Note that use of a non-standard BRI ISDN data combination means that the BRI ISDN equipment will have to be matched to the IDSL, however, 128 kbps should be able to be supported by most consumer BRI ISDN equipment.

Figure 2.7 shows the basic configuration of a BRI ISDN to IDSL connection. Note that, like all xDSL technologies, the physical interface and protocol must be matched at both ends of the physical line. The main difference is that the BRI ISDN equipment does not necessarily expect the protocol to be terminated at the central office. Some loss of BRI ISDN services will certainly happen but it provides a mechanism to use widely available standard BRI ISDN equipment in a potentially lower cost situation.

2.5.4 HDSL/HDSL2

There are two categories for High-speed Digital Subscriber Line (HDSL). As a physical transmission technology, it can be used for general access (same as ADSL), however, it is most popularly used as the physical layer for digital trunk transmissions. In North America and Japan, this transmission type is called T1 and allows transmissions speeds of 1.544 Mbps bidirectionally. In Europe, and many other parts of the world, the E1 standard is used which provides a speed of 2.044 Mbp. Note that specific country usage varies depending on equipment and network choices.

The T1 and E1 framing structure, as shown in Figure 2.8, consist of either 24 or 32 B-channel TDM slots. Each frame is transmitted in 125 microseconds, giving 8,000 frames per second and 64,000 bps for each B-channel. The HDSL frames are transmitted using the same 2B1Q physical coding as the BRI 'U' interface line.

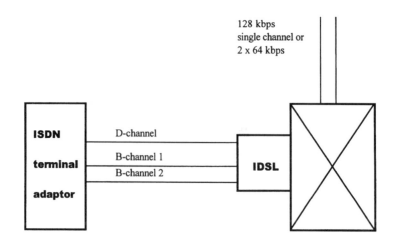

FIGURE 2.7 IDSL configuration.

The xDSL Family of Protocols 39

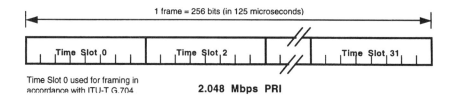

2.048 Mbps PRI

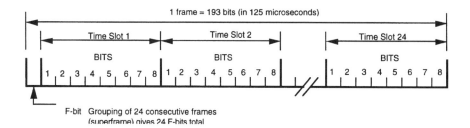

1.544 Mbps PRI

FIGURE 2.8 T1/E1 framing structure for PRI. (Adapted from ITU-T Recommendation I.431.)

However, the transmission speed is multiplied by five, giving a possible usable speed of 800 kbps on a single-twisted pair.

This indicates the real difference between T1/E1 and the underlying physical transmission medium (HDSL or HDSL2 or SHDSL). The framing structure shown in Figure 2.8 (taken from the ITU-T Recommendation) expects a single transmission line and the figure shows a logical frame for either T1 or E1 framing. The real HDSL framing is broken into HDSL frames rather than 64 kbps channels (often referred to as DS-0 channels) plus overhead. As seen in Figure 2.9, each HDSL frame is repeated every 6 milliseconds. Note that this figure indicates that it is impossible to have an integral number of HDSL frames per second (the number is 166 2/3 times per second). HDSL framing adjusts to this by the end "stuffing quats" (four-value quaternary units, as opposed to binary value bits) being determined according to the needs of the overall frame. This means that about every other frame, a 4-bit (2-quat) stuffing unit is added to the HDSL frame.

The HDSL frame, itself, has four logical units. These are the forward overhead units composed of a synchronization unit (14 bits or 7 quats) plus a forward HDSL overhead unit (2 bits or 1 quat), followed by twelve data groups (group 1 or group 2, depending on the twisted wire number—for T1 uses), a group HDSL overhead unit (10 bits or 5 quats), then the twelve data groups/HDSL group overhead repeated twice more, finished up with one more set of twelve data groups and the variable stuffing quats. This gives 4,608 bits of data for each HDSL frame with 94 to 98 bits of overhead (depending on stuffing quat needs). Another way to put it is that an HDSL line can transport 768 kbps of data (or twelve 64 kbps data channels) plus about 16 kbps (96 times 166 2/3 frames per second) overhead.

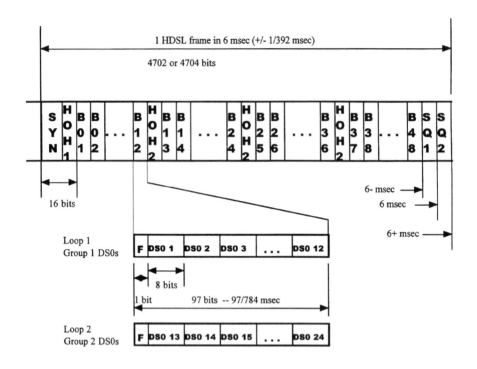

FIGURE 2.9 HDSL DS-1 framing.

We indicated that a T1 frame has a transmission speed of 1.544 Mbps and E1 a speed of 2.044 Mbps. The HDSL technology handles this by using multiple twisted pairs. T1 makes use of two pairs and E1 uses three pairs. This enhances the idea of DSL technology making use of the same twisted pair as conventional analog transmission wiring. By using multiple pairs, greater bandwidth can be achieved. The same is true for analog lines—using two analog lines at a 33.6 kbps speed allows an aggregate of 67.2 kbps. Depending on the specific equipment and tariff structures in a locale, it may be cheaper for the consumer to use two analog lines than to invest in single B-channel BRI ISDN.

2.5.4.1 Signaling Using Channel Associated Signaling

In order to provide access to the PSTN, some type of signaling must be available on an HDSL line (or lines). There are two major flavors. The older networks make use of a signaling form called Channel Associated Signaling (CAS). This means that a bit is associated with a specific B-channel and is used for general signaling purposes over extended periods of time.

For the T1 transmission system, The CAS system is more specifically referred to as "Robbed Bit" signaling. This is because each of the 8-bit time slots for each channel has one bit (the low-order bit) dedicated for signaling purposes. This leaves only 7 bits for data and 7/8 of a 64,000 bps channel leaves 56K data possible (not a coincidence concerning 56K modems). In conjunction with the section on bottlenecks, we can see that use of a Robbed Bit T1 transmission medium in a circuit will thus reduce possible speed to 56K. This is a major difference in ISDN use between North America and Europe.

In Europe, with the E1 transmission system, one of the channels is devoted to signaling, with 1/4 of a bit allocated to each channel for signaling purposes. We say 1/4 because what is done is that it takes four frames to carry the signaling bits for all of the channels. This slows down signaling speeds but means that each of the remaining 30 data channels (remember that one channel of the E1 frame was used for framing) still can carry a full 64,000 bps of data. Thus 64K/56K interworking is not a problem in European ISDNs.

2.5.4.2 Signaling Using Primary Rate Interface ISDN

It is also possible to use the same (slightly varied in content because of physical medium needs) signaling protocols as is used with BRI ISDN. The time slot used for CAS can be used for Q.921/Q.931 message-based signaling instead. A T1 line used with PRI signaling will use the same method as E1—allocating one of the time slots for signaling information. Thus, a T1 PRI line will still give 64,000 bps transmission capability per channel.

This 64-kbps channel controls the signaling needs for 23 (or 30) B-channels. This gives an average of 2.78 (or 2.13) kbps signaling associated with each data channel. BRI ISDN provides 8 kbps (16 kbps D-channel supporting 2 B-channels) per data channel. With this reduced amount of bandwidth available for signaling purposes, the D-channel for PRI is normally dedicated to signaling (no multiplexed data packets supported).

It is also possible for Network Facility Associated Signaling (NFAS) to be supported on a PRI ISDN. This allows a signaling channel on one T1 HDSL to control the data channels on other (associated) T1 HDSLs. This further reduces the signaling bandwidth per data channel, but signaling is really fairly sporadic and the time requirements are pretty slow for the protocols.

2.5.4.3 HDSL2 or SHDSL

HDSL2, or Single-pair High-bit-rate Digital Subscriber Line (SHDSL), technology takes the T1/E1 transmission mapping but seeks to increase the bandwidth on a single-twisted pair to the point that the service can be provided on a single pair. This concept was introduced into the ANSI T1E1.4 group in June 1995. Similar to ADSL, various coding standards have been proposed to achieve these speeds. Also similar to ADSL, the ability to reach these speeds is highly dependent on the condition and length of the local loop. Generally, a distance of 10,000 feet is considered to be the maximum supportable length for SHDSL. Please also note that, although a data group size of 12 DS-0 units is the "normal" HDSL frame definition, by increasing the number (say to 16), it is possible to use only two-twisted pairs for support of E1—or a similar increase in SHDSL. Since the frames are defined to be transmitted per unit time, increasing the number of DS-0 groups per 6 milliseconds will increase the total bits per second transmitted (and also increase the frequency spectrum needed and likely decrease the serviceable length).

2.5.5 SDSL

Sometimes it seems like the acronyms used for xDSL technologies are more in the realm of guesswork than definition. Some references will indicate that SDSL is the same as SHDSL. Others will indicate that it is 1/2 of a T1 using HDSL. In other words, there is no clear unique definition. We will use SDSL as indicating Single-pair Digital Subscriber Line, or a single-twisted pair using the HDSL technology. This provides the same distance limitations as HDSL technology (about 18,000 feet), but making use of only a single twisted pair.

Signaling for SDSL will often be indicated as a "fractional T1" (or E1) and either PRI or CAS signaling may be used, but with a limited number of data-bearer channels.

2.5.6 ADSL/RADSL

Asymmetric Digital Subscription Line (ADSL) and Rate-adaptive Asymmetric Digital Subscription Line (RADSL) can be regarded as the same technology. That is, there is no significant difference on the physical, or protocol, level between the two. The only real difference is that RADSL "calls out" a function often done with both—dynamic changes to the bandwidth (usually only in the "downstream" direction) based on line conditions and other needs.

ADSL, like HDSL, provides a specific physical mechanism for transmission and reception but, does not, in itself, indicate how the service is to be used. The ADSL frame will implicitly define the low-layer use and this will be discussed in detail in Chapter 3. It is not well suited for direct use of Q.921/Q.931 signaling methods, but can be used with Asynchronous Transfer Mode which may have Q.921/Q.931 signaling as part of the protocol.

ADSL equipment primarily follows Carrierless Amplitude/Phase (CAP) modulation, Quadrature Amplitude Modulation (QAM), or Discrete MultiTone (DMT) technology as the coding mechanism for the physical layer. Each relies on separating bands for "upstream" and "downstream" data paths. DMT has been accepted as the primary mechanism in the ANSI T1.413 specification for ADSL. Since ADSL is not well suited as a long-distance technology but must connect into a WAN for wide access, it doesn't much matter which technology is in use as long as both endpoints are compatible.

The xDSL Family of Protocols

Since no signaling channel is implied by the low-level frame structure, it is left to the application to decide just how it needs to connect to the other end of the data path. These options include connection directly into a "nailed-up" (or semipermanent) data line connection (likely T1) to an ISP or other data service provider. Other options bring routers into the central office, so the the data can be distributed to a higher bandwidth LAN, perhaps operated by the switching office. The third set of options carry broadband signaling protocols "piggybacked" onto the ADSL frame structure. ATM signaling is a good possibility for this. The DSL Access Module (DSLAM) existing at the central office, or network, endpoint has the vital requirement to route the data appropriately to and from the ADSL local loop.

With all the available connectivity options, it is extremely important that "classes" of ADSL access units be defined so that equipment can be used with specific services. For example, a specific "class" may be defined for ADSL equipment which is going to connect to a central office provided ISP. Such a class may entail use of TCP/IP within PPP HDLC frames. Another class may require access to the long-distance network and this might mean ATM with AAL 5. These protocols will be covered later in this book. The point, at present, is that different needs require different protocols. BRI and PRI ISDN access methods as well as 56K Modems have specific protocol needs and expectations. The ADSL Forum, in conjunction with other special interest groups, will continue to research these items. The System Network Architecture Group (SNAG) is particularly active in these matters.

2.5.7 CDSL/ADSL "lite"

One of the special features of ADSL is that it is designed to allow the 4-KHz baseband area to remain for use with POTS. This allows "two accesses in one." Although this is also true with BRI ISDN (with POTS ports), it is slightly different with ADSL. In BRI ISDN, the convergence of the POTS signals to the digital access line is performed at the customer's equipment. For ADSL, there are two access groups provided over the same local loop as seen in Figure 2.10.

Since these are two separate access technologies, it is necessary to "split" them before use. The original ADSL specifications basically said that the line would carry both access groups and, prior to entering the central office or residence, be split into the two technologies, as seen in Figure 2.10. This meant that the ADSL equipment would be concerned only with ADSL and the POTS equipment (including switched network) would be solely concerned with the analog voice spectrum.

Keeping the paired technologies together only over the shared medium is a good architecture. However, there were two problems: the main problem is that it required installation of the splitter at the access points for the line (where the local loop entered the building or where it entered the central office). A secondary concern was the need for two different wiring networks (one for existing voice use and a new, separate network for use with ADSL). Both add to total cost for the consumer and the first leads to delays as the operators at the central office must have the remote equipment installed. Delays and added cost both lead to slower deployment of new technologies.

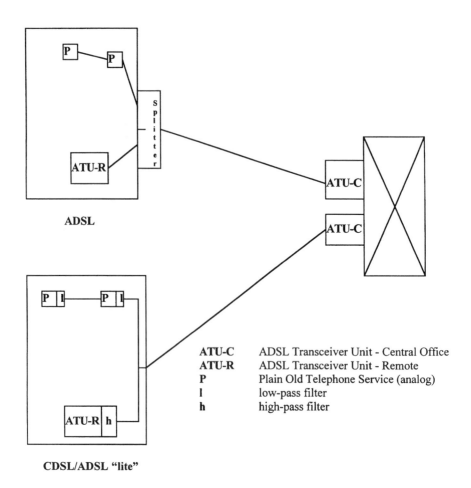

FIGURE 2.10 ADSL and ADSL "lite" line configuration.

Consumer Digital Subscriber Line (CDSL) or ADSL "lite" says basically that it should be possible to make use of the existing wiring within the premise for both ADSL and POTS. This approach does require multiple splitters—low-pass (allowing the baseband 4 KHz to pass through to the equipment) filters on each analog port and high-pass for ADSL equipment. However, new wiring is not needed and the final costs will be based on the equipment the consumer really needs (how many phones, faxes, extensions, ADSL access modems, etc.).

Recalling our discussion about the effect of bridged taps on the ability to carry high-bandwidth data, we should ask whether the architecture affects the ability to carry the ADSL data stream? It does. So, the idea of ADSL "lite" has two components. It allows greater use of existing wiring with less intervention by network personnel but, in turn, it reduces the access speed available over the line. Reduction of line speed also allows the central office to tariff a single service, rather than basing fees on distance from the central office (which affects maximum speed) and line conditions.

In other words, it is possible with full RADSL (using the Rate Adaptive term to emphasize the use of rate adaption) to have different levels of service. Certain neighborhoods might have access to full 4 Mbps downstream speeds while others might only be able to support 800 kbps downstream access. It wouldn't be fair to charge equally for both services and, yet, it's hard with RADSL to pre-determine the level of service before it is installed. CDSL allows a lower, "least common denominator," service to be provided that requires little physical intervention and can provide equal service provisioning.

2.5.8 VDSL

Very high-speed Digital Subscriber Line (VDSL) is considered to be the "next" speed progression in the link from BRI ISDN (to PRI ISDN or SDSL/HDSL) to ADSL to VDSL. ADSL provides greater speed, but requires a new architecture (not yet finalized) to get non-fixed (switched or routed) endpoint access. However, the same lines that can be used for other xDSL services (such as HDSL T1 or BRI ISDN) can be used with ADSL.

VDSL requires the limit on length to be reduced, probably to a limit of only 1,000 to 2,000 feet (0.3 to 0.6 kilometers) and the speed increased to 30 to 50 Mbps. Use of VDSL technology is expected to be tightly linked to the deployment of Fiber-To-The-Curb (FTTC) because, in order for it to be practical for many VDSL units to be in use, a "hub" system is likely to be used. These hub units will shift the local loop to concentrated bundles of fiber optics which will then lead into the switched, or routed, or redirected high-speed links.

This step says that the current infrastructure will be slowly replaced starting from the long-distance networks and making its way to the final residence or business. The long-distance networks are largely high-speed ones but T1/E1 (and higher capacity bundles labeled T3) copper-based trunks are still in dominant use. The long-distance network for new services will need to have gigabit transmission capabilities which implies a probable fiber optic situation. Bringing that fiber optic linkage to the neighborhood hubs will allow only the existing wiring within homes and residences to remain the same. Use of VDSL and possible migration strategies will be discussed in greater detail in Chapter 10.

2.6 SUMMARY OF THE xDSL FAMILY

A wide deployment of xDSL technologies already exists if we consider the digital influence on existing higher speed analog modems. These are actually hybrid technologies. The BRI and PRI ISDN access methods allow digital replacement for entry into the long-distance network. This provides direct Wide Area Network (WAN) system control and information passage as well as maximizing the possibility of using the digital data channel capacities. HDSL and HDSL2/SHDSL provide a framing mechanism to help support these technologies. SDSL gives medium-speed access possibility without needing to change use of single-twisted pairs.

ADSL provides the access method to start the separation of equipment use from the existing infrastructure, but retains most of the local loop intact. VDSL requires replacement of the infrastructure and migration of such closer to the neighborhoods. Thus, in total, xDSL technologies provide a migration strategy from the networks of the 1800s to the desired systems of the 2000s.

3 The ADSL Physical Layer Protocol

The Asymmetric Digital Subscriber Line (ADSL) technology was originally devised in the laboratory to solve "Video on Demand" requirements. (A 1.5 Mbps data stream is sufficient for MPEG-I video streams, although about 8 Mbps is used for MPEG-II and DVD/HDTV.) However, since then, the Internet has probably developed into the primary application for the service. Originally designed to provide up to 8 Mbps in the downstream direction (from the central office or data service provider to the user equipment) and 64 to 128 kbps in the upstream direction, various experimental deployments have indicated that this is not practical for most real-life local loops.

The development of ADSL, and most of the xDSL technologies, has been highly dependent on the ability to have very complex adaptive circuitry utilizing Very-Large-Scale Integrated circuit (VLSI). This allows very fast processing of data units *and* the ability to change algorithms and organization rapidly based on changing line conditions.

It was possible to use 2B1Q line coding for ADSL, in fact, SDSL is almost the same transmission speed as that decided on for CDSL. However, there are variations in line conditions that make a 2B1Q approach less sturdy. Instead, mechanisms which allowed the frequency spectrum to be easily broken into subunits were investigated—very useful for rate adaptive situations. These were Quadrature Amplitude Modulation (QAM), Carrierless Amplitude/Phase modulation (CAP), and Discrete MultiTone (DMT). In each case, there is the ability to transmit a higher density of information per cycle by using 3D coding or "chord"—multiple single codes—systems.

3.1 CAP/QAM

CAP is actually a subset of Quadrature Amplitude Modulation (QAM). QAM makes use of three dimensions to provide values. These dimensions are amplitude, phase, and frequency. Generally, one of these dimensions—frequency—is held at a constant value while the amplitude and phase are modulated (as in the AP part of CAP).

In this type of coding system there are two separate signals generated (three for non-carrierless): a sine wave and a cosine wave. Normally, the sine wave and cosine wave are 90° apart, as shown in Figure 3.1. The sine wave and/or cosine wave can be phase shifted within the same frequency. It is now possible that the sine wave and cosine can be 180° apart (one is starting the positive portion while the other is starting to go negative), or phase-shifted to any other relationship. Each wave can

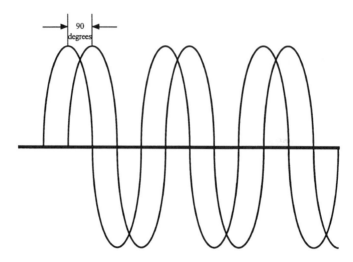

FIGURE 3.1 Sine and cosine wave forms.

also have multiple amplitudes. So, it would be possible for each sine/cosine wave to be in one of four different phase positions and each wave to have one of two different amplitudes. These four phase positions times two amplitudes (sine) times two amplitudes (cosine) gives the possibility, as listed in Table 3.1, of 16 separate, distinct, values. Note that these are not necessarily the values used for the specific QAM system in use by the equipment—it is only an example of a QAM system.

The amount of phase offsetting is really only possible to look at from the point of view of an "ideal" sine or cosine wave. Thus, with a general QAM encoding/decoding system, it is possible to send an unmodulated carrier along with the signal as a reference. However, if it is a carrierless system, then the original baseline signal is superimposed by logic channels (no physical carrier is present but a "pseudo-carrier" can be inferred).

It is also possible for a 16 QAM to be generated by keeping the sine/cosine phase shift locked (say at the "normal" 90° separation) and have each sine/cosine wave have one of four values. Thus, with four possible amplitudes (sine) times four amplitudes (cosine), we once again have sixteen possible values for each wave cycle.

TABLE 3.1
Example 16-Value Phase/Amplitude Chart

	Sine/Cosine			
Phase	Low/Low	Low/High	High/Low	High/High
0°	0000	0001	0010	0011
90°	0100	0101	0110	0111
180°	1000	1001	1010	1011
270°	1100	1101	1110	1111

The ADSL Physical Layer Protocol

Carrierless Amplitude/Phase Modulation is a variant of QAM that does not have a specific carrier wave set up for the baseline comparison. It is sometimes referred to as "carrier suppressed."

The matrix associated with a QAM system may also be referred to as a "constellation." This is because there is no requirement for all values to be used, which means that it is not necessarily able to be put into a simple matrix form, or. if so, into a sparse matrix.

3.2 DISCRETE MULTITONE

Discrete MultiTone (DMT) is often discussed with CAP as being a conflicting technology. In actuality, DMT makes use of variations of the QAM/CAP coding method. However, the primary technical difference, pioneered at Bell Laboratories, is the idea of breaking up the frequency spectrum into equally spaced subchannels. Sometimes these are considered to be *subcarriers,* meaning that different carrier waves (explicit or suppressed/extrapolated) are used to base the coding for each subarea of the frequency spectrum.

The spectrum, considered reasonably usable, extends from the base (0 Hz) to about 1.1 MHz. By dividing the spectrum evenly into 4.3125 KHz bands, it is possible to have 256 subchannels available for information—channels 1 through 256. (Note: when doing the arithmetic, it is discovered that 1.104 MHz is actually used.) Splitting the spectrum does not add any information-handling capability. What it does do is allow for interference and "automatic" rate adaption by eliminating subchannels from a data path (upstream or downstream—although it is most likely to be a downstream subchannel).

Let's say that each 4-KHz subchannel can support 64 kbps of data (4,000 cycles per second with 16 values per cycle using 16 QAM). The 256 subchannels can then provide a theoretical capacity of greater than 16 Mbps. In actuality, such speeds are not often feasible (except in the laboratory or tightly controlled situations. First, many coding schemes only do an 8 QAM, reducing the possible bandwidth to 8 Mbps. Next, it is important to keep the low baseband of 0 to 4 KHz free for voice. To prevent "seepage" from the ADSL frequencies to the speech bands, subchannels 1 through 6 are often reserved to preserve a "guardband" between the active channel for speech and the first active channel for ADSL.

Whenever the same frequency is used for bidirectional traffic (such as speech), some type of echo situation is going to occur. This happens when the transmitted signal is reflected ("echoed") back the same as the originating direction. There are various echo cancellers which basically "subtract" the diminished form of the originating signal after certain delay times. If a person says "HELLO" into a canyon, then they must subtract the "hello" that comes back a bit later to understand properly what someone else is saying at that time.

However, echo cancellation becomes a mostly moot point (there can still be problems with cascading echoes causing signal degradation) if different subchannels are used for different directions. Thus, the UTP becomes three access points: one for speech, one for data from the user to the central office ("upstream"), and one for data from the central office to the user ("downstream"). Since the channels are fully separated, there is little problem with echoes.

3.3 ANSI T1.413

Until the ITU-T Recommendations are released later in 1999, the ANSI T1.413 specification is the primary public document for ADSL. Note, however, that this isn't necessarily a bad point. First, since so much extensive testing, and experimental work has been done with ADSL based on work incorporated into the ANSI document, the ITU-T Recommendation(s) are likely to be very similar to those already published. Second, the ADSL "lite" requirements will be "less than" types of requirements. In other words, certain possibilities and services will be eliminated, maximum allowable data rates will be reduced, and required equipment at the customer's premises will be avoided. Study of ANSI T1.413 as the base document is, therefore, very reasonable. Issue 1 has been used as the implementation base for a wide variety of equipment and, although de-emphasized, Issue 2 still allows equivalent equipment to be designed and implemented.

Issue 2 of ANSI T1.413 is currently in draft form and, therefore, may still be subject to change. There are quite a few changes between Issue 1 and Issue 2 (more than are listed in Annex N of Issue 2). The main differences are a de-emphasis on unstructured use of ADSL (now referred to as being part of Synchronous Transfer Mode, STM) and a greater emphasis on ATM. Probably in preparation for CDSL, a greater flexibility has been added for STM ADSL such that it is more flexible in bearer capacities (see the next section). Where not confusing, both Issue 1 and Issue 2 aspects of ADSL will be discussed.

The major portions of T1.413 (Issue 1 and 2) will be discussed in this section, but the details will not attempt to duplicate the contents of the ANSI document. In the first place, a full expansion of the document would fill a large book and would really serve no useful function; examining the most recent ANSI specifications will be more accurate and concise. Another thing, however, is that most specific electrical manipulations must be done by some type of semiconductor device. This will be examined in more detail in Chapter 5. As mentioned at the beginning of this chapter, only VLSI technology has enabled the possibility of high-speed, error-correctable transmission line technologies. It is necessary to support the speed of the line and to be able to react appropriately to changing line conditions and needs. This section will explain the general purposes and structure of the ADSL superframe, frames, and special bytes to be used as a quick and direct reference against the interfaces supplied by the semiconductor chip manufacturers.

The document T1.413 is titled "Network and Customer Installation Interfaces—Asymmetric Digital Subscriber Line (ADSL) Metallic Interface." In other words, the document is intended for both central office and user equipment and is oriented toward copper electrical circuits. The general system reference model, seen in Figure 3.2, gives the general functional blocks needed to provide ADSL service. The difference between this diagram and CDSL is primarily the removal of the splitter (at least at the ADSL Transceiver Unit-Remote terminal end [ATU-R] with the requirement of a low-pass filter before any POTS equipment attached to the line

The ADSL Physical Layer Protocol

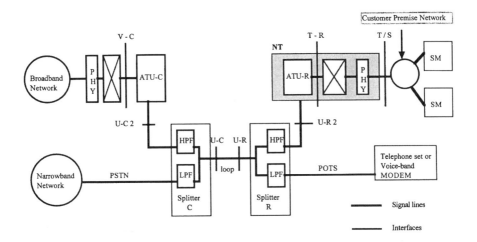

FIGURE 3.2 ADSL system reference model. (From ANSI T1.413.)

and a high-pass filter before the ATU-R. (The high-pass filter may be incorporated into the ATU-R equipment, but the low-pass filter for POTS equipment is likely to be a separate box.)

Figure 3.3 shows the ADSL Transceiver Unit-Central office end (ATU-C). The ATU-R (Remote end) looks the same except that only the three Lsx channels are going upstream and x_n and Z_i variable ranges change. Issue 2 adds a Network Timing Reference (NTR) input on the ATU-C side—for what is now called the STM transport reference model. Note that there are only two differences. The ATU-R transmits only on the LSx "duplex" subchannels while the ATU-C potentially transmits on both the ASx simplex channels *and* LSx duplex channels. Also, the ATU-R is limited to the first 32 (k) subchannels (actually, probably both sides will be limited starting at higher than 0 for the lower order) and the ATU-C has access to all 256 subchannels. Otherwise, both are the same. Basically, the ATU-C may transmit more data to the ATU-R than vice versa.

Figure 3.4 shows the ATU-C transmitter reference model for ATM transport. (ATU-R Remote end has similar differences to the ATU-C as for the STM model.) The main differences between the STM and ATM models is that only AS0 (and optionally AS1) are used for downstream transmission with the ATU-C. Each ATMx transport line passes through a Cell Transmission Convergence (TC) layer before being passed through the AS0 or AS1 bearer channel. On the ATU-R side, the ATMx streams proceed through the LS0 or LS1 bearer channels.

3.3.1 BEARER CHANNELS

ADSL has the same concept of bearer channels; however, they are broken down into two parts. These are "simplex" and "duplex" channels allowing for multiple upstream and downstream channels. In what Issue 2 calls the STM mode, there are four

52 ADSL: Standards, Implementation, and Architecture

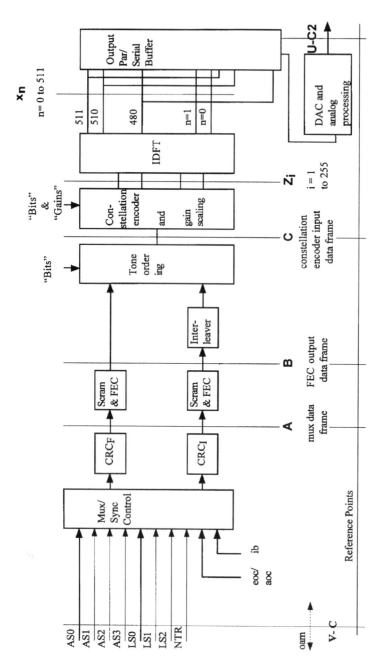

FIGURE 3.3 ATU-C transmitter reference model for STM transport. (From ANSI T1.413.)

The ADSL Physical Layer Protocol

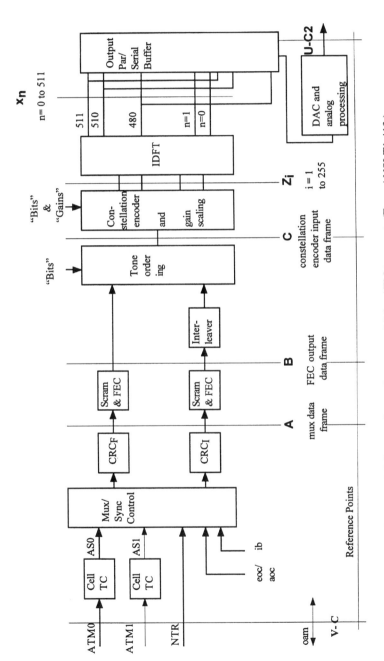

FIGURE 3.4 ATU-C transmitter reference model for ATM transport. (From ANSI T1.413.)

possible simplex bearer channels (for NA applications), called "ASx" that are divided into multiples of 1.536 Mbps.

Issue 1 of T1.413 discusses the aggregate data traffic according to "transport class." Issue 2 keeps the general ideas of the same ranges but considers Issue 1's transport classes as "maximum" values with a minimum of a single DMT subchannel (32 kbps) as possible.

Transport class 1 has a total aggregate data rate of 6.144 Mbps. Class 2 has 4.608 Mbps. Class 3 has 3.072 Mbps and class 4 has 1.536 Mbps capacity. These classes are basically arranged according to shortest distance/highest capacity to longest distance/lowest capacity. The ASx (AS0, AS1, AS2, and AS3) duplex bearer channels can have a multiple of 1.536 Mbps per channel with the total depending on the transport class. AS0 at transport class 1 can have 6.144 Mbps with AS1-AS3 having no capacity. Or AS0 might have 3.072 Mbps, AS1 have 1.536 Mbps and AS2 have the last 1.536 Mbps available. Table 3.2 gives a summary of possibilities.

The same type of categorization happens with European/E1 types of lines. The total possible aggregate is still 6.144 Mbps. However, transport classes 2M-1, 2M-2, and 2M-3 have possible totals of 6.144 Mbps, 4.096 Mbps, and 2.048 Mbps respectively and there are only AS0-AS2 subchannel designations. T1.413 further breaks this possible throughput into Asynchronous Transport Mode (ATM, a broad-

TABLE 3.2
Bearer Channel Options by Transport Class for Bearer Rates Based on Downstream Multiples of 1.536 Mbit/s.

Transport Class:	1	2	3	4
Downstream Simplex Bearers:				
Maximum capacity	6.144 Mbps	4.608 Mbps	3.072 Mbps	1.536 Mbps
Bearer channel options	1.536 Mbps, 3.072 Mbps, 4.608 Mbps, 6.144 Mbps	1.536 Mbps, 3.072 Mbps, 4.608 Mbps	1.536 Mbps, 3.072 Mbps	1.536 Mbps
Maximum active subchannels	4 (AS0, AS1, AS2, AS3)	3 (AS0, AS1, AS2)	2 (AS0, AS1)	1 (AS0 only)
Duplex Bearers:				
Maximum capacity	640 kbps	608 kbps	608 kbps	176 kbps
Bearer channel options	576 kbps, 384 kbps, 160 kbps, C (64 kbps)	576 kbps,* 386 kbps, 160 kbps, C (64 kbps)	576 kbp,* 384 kbps, 160 kbps, C (64 kbps)	160 kbps, C (16 kbps)
Maximum active subchannels	3 (LS0, LS1, LS2)	2 (LS0, LS1,) or (LS0, LS2)	2 (LS0, LS1) or (LS0, LS2)	2 (LS0, LS1)

* For further study.

Source: From ANSI T1.413, Issue 1.

The ADSL Physical Layer Protocol   55

band ISDN system, will be discussed in greater depth in Chapters 6 and 7) data cell bit rates.

The second category of bearer sub-channel is the "duplex bearers." The possible sub-channels on these are the C Channel, LS0, LS1, and LS2 channels. The C (Control) Channel is mandatory and, for transport class 4, is carried (at 16 kbps) within the ADSL synchronization overhead. For other classes, it is carried over the LS0 channel at 64 kbps. LS1 and LS2 are therefore parallel in purpose and distribution to the ASx channels. The total bandwidth available ranges from 640 kbps to 608 kbps to 176 kbps. Issue 1 states that LS1 may be used for 160 kbps channel and LS2 used for 384 kbps (or 576 kbps if LS1 is not in use). Note that this 160 kbps is the size for a BRI ISDN frame (without the 'U' or 'S'/'T' frame overhead). Issue 1 called this out as to be used for transport of BRI ISDN transport; issue 2 deletes this as an explicit service. European transport classes basically merge NA transport classes 2 and 3 into the 2M-2 category.

Asynchronous mode (such as ATM) basically says that there is no fixed throughput speed—only a maximum one. However, in both the STM and ATM cases, this is achieved by using a fixed bit rate but with "idle" or "fill" bytes put into the data stream. This keeps the bit-rate constant (allowing for synchronization of superframes) but allows the real data rate to adjust as needed. This is rather like using a large box for shipping a small item and then putting in padding to fill the box—only the data is truly useful but it may fill any amount of the volume within the box.

Issue 2 indicates that if a single latency channel is being used (see next section) then only AS0 and LS0 are to be used (with the ATU-C using an incoming NTR). If both latency channels are being used, then AS0/LS0 is assigned to one latency channel type and AS1/LS1 is assigned to the other latency channel type. Each subchannel shall be a multiple of 32 kbps, with a maximum of 6.144 Mbps in the downstream direction (ASx) and 640 kbps in the upstream direction (LSx).

3.3.2 ADSL Superframe Structure

Issue 2 defines four different framing structures: 0, 1, 2, and 3. Frame structure 0 is basically what was defined in Issue 1. Frame structure 1 disables the synchronization control mechanism for use with synchronous frames. Framing structures 2 and 3 provide reduced overhead framing with either separate or merged fast and sync bytes. The ATU-C must support all "lower" numbered frame structures in addition to the highest (for example, if it supports framing structure 2, it *must* also support framing structures 0 and 1). The ATU-R has primary control over the final framing structure chosen for the link.

The superframe for ADSL is shown in Figure 3.5. A superframe is basically the large envelope for ADSL—a collection of smaller frames. Each superframe, consisting of 68 frames plus a synchronization frame, is transmitted in 17 milliseconds. Since the DMT frequency carrier works with 4 KHz (actually 4.3125, but the band is not used to the "edges"), each frame must be transmitted in 250 microseconds. Since the synchronization frame is not actually sent, it is overhead (1 frame out of 69 total), and this overhead must be allowed for by sending each data frame buffer within 68/69 times 250 microseconds.

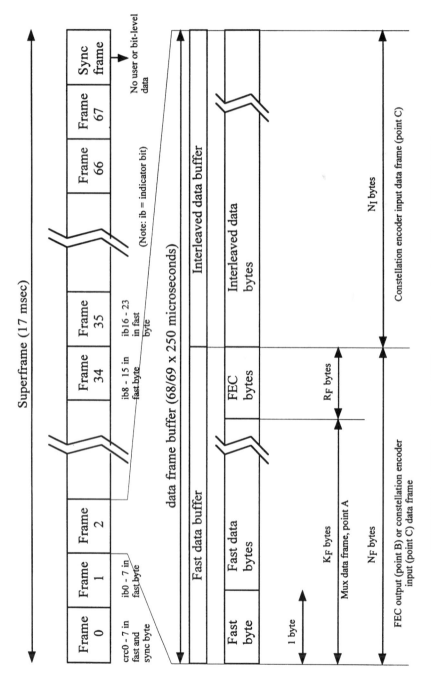

FIGURE 3.5 ATU-C transmitter ADSL superframe structure. (From ANSI T1.413.)

3.3.2.1 Fast Data and Interleaved Data

The first area to discuss concerning the ADSL frame is the concept of fast data and interleaved data. Interleaved data undergoes a separate process of "weaving" the data cells together in order to reduce noise. This is done by dispersing the bits across different transmission bands so it is less likely that a group of consecutive bits will be in error—improving the odds that the Forward Error Correction (FEC) can determine the error and "autocorrect" it.

There are two main methods for accomplishing block interleaving and convolutional interleaving. Both types of interleaving change the order of the transmitted bits of an outgoing stream, but do so in a formulaic fashion so that they can be re-sorted upon receipt. Block interleaving fills in a fixed size "block" by rows and then sends the data out by "columns" as in Figure 3.6. Convolutional interleaving offsets the block (trapezoidal) so that a circular buffer can be used with the rows written to and columns read out in a similar fashion to block interleaving but, with the offsetting and circular buffering, on a more constant basis.

Fast data is just non-interleaved data. Interleaving helps to reduce noise errors, but the process of interleaving and (on receipt) de-interleaving increases the latency period associated with the data. So, fast data is really low-latency data with a greater (compared to interleaved data) susceptibility to noise. Interleaved data have longer latency, but better protection against noise. Video on Demand is an application well suited to interleaving—it doesn't matter if it takes an extra second or two to begin if the quality of the transmission stays high. An example of data well suited to fast data would be control data (such as call setup information).

When the ADSL is being used for an ATM application, only one data stream is involved (AS0 and LS0) *unless* there is need for both fast and interleaved data. In this case, AS0 and AS1 (and possibly LS0 and LS1) are used, with fast data using one transmission route and interleaved data using the other. This facilitates keeping the applications, which are best suited for each transmission type, synchronized with a particular path.

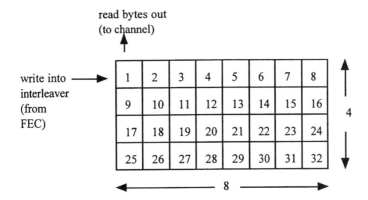

Order of bytes read out to the channel will be: 1, 9, 17, 25, 2, 10, 18, 26, 3, ...

FIGURE 3.6 Block interleaver with a span of 8 and depth of 4.

3.3.2.2 Fast byte

The "fast byte" or first byte in the fast data buffer is used for multiple purposes, as seen in Figure 3.7. These general purposes are for Cyclic Redundancy Check (CRC) of the superframe (CRC), Indicator Bit (ib), Embedded Operations Control (eoc), and sync control (sc) bits. In frames 2 through 33 and 36 through 67, eoc versus sc bits are determined by the value of the least significant bit of the "fast byte." Note also, however, that these are the areas in which Issue 2 has determined that a savings of overhead may be obtained. When they are not required for synchronization control, crc or as ib, an even frame/odd frame set of fast bytes may be used to transmit an eoc message consisting of 13 bits. They may also just contain indications of "no synchronization action."

Saving bytes within the fast byte, or sync byte, doesn't make a lot of sense within the STM world. This is because it is more likely that multiple ASx channels will be in use as well as a number of LSx channels. If the channels in use increase, the overhead decreases. However, with ATM use, the number of channels will decrease (to no more than four total) and some of the synchronization activities become less useful and it becomes overhead that does not provide enough use that it can be safely discarded.

So, there are two basic forms of reducing overhead, primarily used when there are only single channels in each direction, or secondarily, when only a single fast channel is in use and a single interleaved channel is used. While this is particularly defined for use with ATM, it can also be used for an STM application that only needs single channels. Note that the word "channels" is used in a variety of senses. There are the channels (or subchannels, or subcarrier bands) that are individual DMT bands from the potential 256 bands active. There are the channels (ASx and LSx) which are used for application conduits for the data. Then there are the channels used within the transported protocol layers. Although they each have different names, the name channel may often be used for them and the specific purpose is determined from context.

3.3.2.3 Sync Byte and SC Bits

In full overhead mode, the sync byte is used in conjunction with the fast byte to provide a full set of information about the frame. In reduced overhead modes, the fast byte is used for fast buffers and the sync byte is used for interleaved data buffers. Table 3.3 lists overhead functions for reduced overhead mode. The sync bytes (or bytes, when the fast byte is used with its "sc" bits on frame 67) act as a mechanism to designate the ASx and LSx channels being used for the fast or interleaved data buffer and to allow alterations to the data according to the AEX and LEX bytes. In full overhead mode, the sync byte is used for interleaved data on frames 1 through 67 and on the fast data for frame 67.

3.3.2.4 Indicator Bits

Indicator bits are used for notifications. These include detections of errors, corrected errors, loss of signal, remote defects and other information that may either be desired

The ADSL Physical Layer Protocol

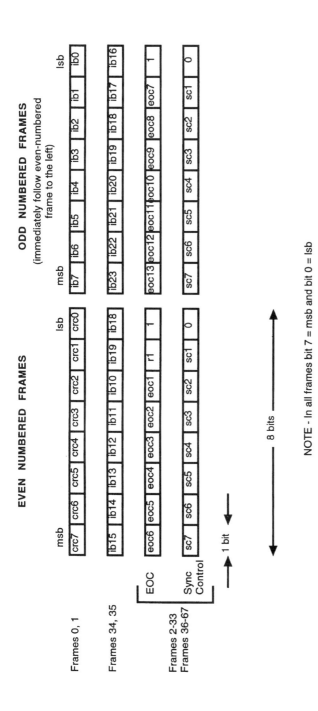

FIGURE 3.7 Fast sync byte ("fast" byte) format—ATU-C transmitter. (From ANSI T1.413.)

TABLE 3.3
Overhead Functions for Reduced Overhead Mode with Merged Fast and Sync Bytes

Frame Number	(Fast Buffer Only) Fast Byte Format	(Interleaved Buffer Only) Sync Byte Format
0	fast CRC	Interleaved CRC
1	ib0 –7	ib0 – 7
34	ib8 – 15	ib8 – 15
35	ib16 – 23	ib16 – 23
4n+2, 4n+3 with n = 0...16, n != 8	sync or eoc*	sync or eoc*
4n, 4n+1 with n = 0,...16, n != 0	aoc	aoc

*In reduced overhead mode only the "no synchronization action" code shall be used.

Source: From ANSI T1.413.

for statistics or for possible higher level correction. Fast buffer indicator bits (downstream direction) are used for far end block errors, far end correction counts, loss of signal and remote defects.

3.3.2.5 CRC Bits

Cyclic Redundancy Check (CRC) bytes are fairly direct in purpose. Basically, the bits of the data stream are used within a mathematical formula so that a summary indication indicates what the previous bytes have been. Naturally, with only 8 bits available, bytes and potential errors cannot be identified to the last bit; however, the errors can be classified into 256 categories. If there are fewer than 256 bytes in the frame to which the CRC is attached, some error correction can possibly take place. In many HDLC frames, a 2-byte CRC (or CRC-16) is used, allowing errors to be categorized into one of 64K possible categories. However, HDLC rarely attempts to do error correction from use of the CRC-16 trailer.

3.3.3 EMBEDDED OPERATIONS CONTROL

The crc, ib, and sc bits are all used for configuration and error information. The eoc bits provide the "real" programming interface that allows change within the ADSL configuration. Table 3.4 shows the bytes of the eoc frame structure. The first two bits indicates whether the information or command is addressed to the ATU-C (value "11") or ATU-R (value "00"). Note that *replies* are not the same as *commands*. If an ATU-C sends a command to the ATU-R, it will put the ATU-R (value "00") in as the address field but, when the ATU-C replies to the command, it will put its own address (ATU-R, value "00") in the address field to indicate that the reply is coming from an ATU-R.

The next bit (eoc3) indicates whether the "information field" (eoc bits 6 through 13) is used for a command or data. Since only one byte of data (or command) is

TABLE 3.4
eoc Message Fields

Field #	Bit(s)	Description	Notes
1	1–2	Address field	Can address 4 locations
2	3	Data (0) or opcode (1) field	Data used for read/write
3	4	Byte parity field Odd (1) or even (0)	Byte order indication for multibyte transmission
4	5	Message/Response field Message/Response message (1) or Autonomous message (0)	Currently no autonomous messages are defined for the ATU-C; the "dying gasp" message is the only autonomous message defined for the ATU-R
5	6–13	Information field	One out of 58 opcodes or 8 bits of data

Source: From ANSI T1.413.

allowed at a time within the eoc bits, it is important to be able to make sure that a nominal sequencing takes place. Bit eoc4 allows for indication of "odd" value ("1") or even ("0") bytes within a character data stream.

Bit eoc5 is rather specialized. It indicates whether the message is "autonomous" (value "0") or not. The only time this is set for autonomous, is for the ATU-R when it wants to indicate that it has lost power (sending these last messages on capacitor or battery backup) and indicates the "dying gasp" towards the ATU-C. So, the only real message in this class has a fixed value. This is sent by the ATU-R to the ATU-C at least six times and, during this time, other eoc commands from the ATU-C shall be ignored.

There are three types of eoc messages: bidirectional messages are originally sent by the ATU-C to the ATU-R and then echoed by the ATU-R back to the ATU-C, ATU-C to ATU-R (downstream) messages, and ATU-R to ATU-C response messages and the autonomous "dying gasp" message. Table 3.5 gives a list of messages, as presently indicated in Issue 2.

Some commands are oriented toward changing ATU-R circuitry. Others obtain statistical information from the ATU-R. Some commands "latch," that is, they cause the ATU-R to stay in the prescribed mode until a further command causes it to change (either a "nonlatched" command or a "back to normal" command). *Request corrupt crc* (for test purposes) and *notify of corrupt crc* (once again, for test purposes) are latching commands. Some messages cause the ATU-R to return data information. Messages can basically be considered to be test messages (send corrupted crc, notify that corrupted crc will be sent, self test, etc.), data transfer messages (mainly of register information), and vendor specific messages (four of which are defined in the ADSL eoc message specs). Note that commands are repeated in order to be certain that they have been received, inasmuch as there is not always a specific acknowledgment.

TABLE 3.5
eoc Message Opcodes

Hex	Opcode Meaning	Direction	Abbreviations/Notes
01	Hold state	d/u	HOLD
F0	Return all active conditions to normal	d/u	RTN
02	Perform "self test"	d/u	SLFTST
04	Unable to Comply (UTC)	u	UTC
07	Request corrupt CRC	d/u	REQCOR (latching)
08	Request end of corrupt CRC	d/u	REQEND
0B	Notify corrupt crc	d/u	NOTCOR (latching)
0D	Notify end of corrupt CRC	d/u	NOTEND
0E	End of data	d/u	EOD
10	Next byte	d	NEXT
13	Request test parameters update	d/u	REQTPU
(20, 23, 25, 26, 29, 2A, 2C, 2F, 31, 32, 34, 37, 38, 3B, 3D, 3E)	Write data register numbers 0–F	d/u	WRITE
(40, 43, 45, 46, 49, 4A, 4C, 4F, 51, 52, 54, 57, 58, 5B, 5D, 5E)	Read data register numbers 0–F	d/u	READ
(19, 1A, 1C, 1F)	Vendor proprietary protocols	d/u	Four opcodes are reserved for vendor proprietary use
E7	Dying gasp	u	DGASP
(15, 16, 80, 83, 85, 86, 89, 8A, 8C, 8F)	Undefined codes		These codes are reserved for future use and shall not be used for any purpose

*d = down; u = up.

Source: From ANSI T1.413.

3.4 ADSL "lite"

The draft ITU-T Recommendation G.922.2 is titled "Splitterless Asymmetric Digital Subscriber Line (ADSL) Transceivers." Since it is a draft (not scheduled to be voted upon until June of 1999), some items may change. However, the basic features are likely to remain, so a short discussion of differences is helpful.

The first important issue is the "splitterless" item. As mentioned before, this means that the POTS (or BRI ISDN) which is carried over the local loop is not split

The ADSL Physical Layer Protocol

off from the ADSL spectrum at the point where it enters the residence or business. This means that no special equipment needs to be installed at the end of the local loop—simplifying installation and reducing the cost. However, it also means that any currently existing POTS or BRI ISDN equipment on the line must use low-pass filters to eliminate the ADSL signal before processing the data.

The next item, is that G.922.2 specifies that ATM will be used. Only AS0 and LS0 channels will be made available. Each will be of a multiple of 32 kbps with the range for the AS0 channel to be from 64 kbps to 1.536 Mbps and the range for the LS0 to be from 32 kbps to 512 kbps.

Next, the headers and frames have been simplified considerably. The frame is equivalent to the "reduced overhead mode with merged fast and sync bytes" using only the "interleave buffer" definition. This supports only a *simplex* AS0 channel downstream and a *simplex* LS0 upstream channel. Thus the data frame consists of a fast/sync bytes and a set of bytes associated with the channel (AS0 downstream and LS0 upstream).

The use of the merged fast/sync byte is shown in Table 3.6. Note that the EOC bytes have the first 6 bits (eoc1 through eoc6) in the even numbered data frames and the upper-order 7 bits (eoc7 through eoc13) in the odd numbered data frames. Some Indicator bits are not used for ADLS "lite," namely, ib10, ib11, ib15, and ib17 as these are all indicator bits applying to the fast data path.

The Autonomous message field is used a bit differently than for full ADSL (T1.413). The bit is set to '1' for acknowledged data transfers from the ATU-C to the ATU-R. The ATU-R should set the bit to '1' for responses to such commands from the ATU-C. The ATU-C should set the bit to '0' for autonomous data transfers. If the ATU-R wants to do an autonomous transfer, it should set the bit to '0'.

The ADSL "lite" eoc message set has a minimal change from that listed in T1.413. Mainly, a new command from the ATU-R to ATU-C is available. This is called REQPDN (value 15) to initiate a transition to a new power management link state. The ATU-C may respond with a GNTPDN (value 16) to grant the transition (and sends the allowed new state) or it may reject the request with a REJPDN (value 83).

TABLE 3.6
Use of Sync Byte in ADSL "lite"

Data Frame (DF)	Sync Byte Contents
0	CRC
1	ib0 - 7
34	ib8 - 15
35	ib16 - 23
4n+2, 4n+3 with n = 0...16, n != 8	eoc
4n, 4n+1 with n = 1...16	aoc

Source: From ITU-T Draft Recommendation G.992.2.

3.5 ATU-R VERSUS ATU-C

As can be seen in the T1.413 section, the ATU-C is primarily in control of the connection between units (although, as mentioned in Issue 2, the ATU-R controls what frame types are actually in use). Except for a few instances (with current standards), the ATU-R is in the position of responding to commands from the ATU-C. This is similar to other devices, such as BRI ISDN, on the 'U' interface where the network is connected to a consumer premise equipment.

The important part of the equipment is that they be matched. If the ATU-R unit makes use of CAP coding techniques then so must the ATU-C unit. This doesn't mean that the ATU-C unit may not be a superset of capabilities. It may not be economically viable for an ATU-C to be able to do every option possible. However, it will make sense for the equipment on the network side to be more versatile than a specific ATU-R. Much of the content issues are DSLAM (DSL Access Multiplexer) problems as the specific data contents are the duty of what the ATU-R or ATU-C is hooked to.

So the ATU-C and ATU-R are a matched set, but not identical. The ATU-R can query and command the ATU-R. The ATU-R provides the configuration abilities for the line. The ATU-R may be connected into a LAN or it may provide some direct access to host applications. The ATU-C must provide the connectivity from ADSL (which *is* only useful for the link between the consumer premises to the network access) to whatever service is desired.

3.6 DSLAM COMPONENTS

The DSL Access Multiplexer is a unit, separate from the ATU-C, which provides access to a multitude of potential LANs, WANs, and services. A general architecture of a DSLAM is shown in Figure 3.8. The word *multiple*, which is effectively part of the word multiplexer, summarizes the DSLAM.

The DSLAM architecture is composed of three basic components. The first set is made up of the subscriber links, which is the network-side of various subscriber access methods. Within this book, the ATU-R would certainly be the "most important." However, within the concept of a DSLAM, there is no"most-important" access method. There can be devices using any of the xDSL technologies—56K Modems, ADSL/RADSL, HDSL, SDSL, VDSL—or even POTS lines.

What is fed into the DSLAM is not fully agreed upon. Some books and references indicate that a full, unmodified, bit stream from the service device would be brought into the DSLAM. However, to a certain extent, this violates some of the symmetricity of devices. A stream of data is fed into an ATU-R. This might be ATM cells or it might be a bit transfer stream (STM) or it might be a series of packets. The ATU-R will take these input streams and place them into the appropriate CAP or DMT format. As part of this process, to ensure constant bit rates, some insertion of idle bits or fillers may be necessary. These will arrive at the ATU-R. It makes sense that the ATU-R will then *remove* any inserted "idle" bits or otherwise unusable "non-data." This means that ADSL (and most other xDSL technologies) input into the

The ADSL Physical Layer Protocol

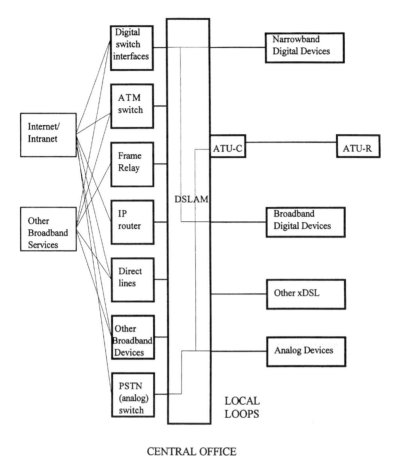

FIGURE 3.8 DSL Access multiplexer description.

DSLAM will provide only a maximum burst level at the speed officially supported by the customer device.

On the other side of the DSLAM are the various network services, or connections, that allow the subscribers access to desired endpoints. These may be access to the PSTN, in which case, it is quite likely that the DSLAM will provide "clear channel" access between the POTS channel of the ATU-C or a direct 56K Modem analog link. This saves on unnecessary complexity within the DSLAM for it to duplicate.

Beyond POTS, which is a very simple potential provision of a DSLAM, there are routers, WANs, and "direct" connections to ISPs. Routers envisioned for the DSLAM are primarily of the Internet Protocol (IP) type. However, it is also possible for the DSLAM to connect into a Frame Relay router network. In these cases, the data streams coming in from the subscriber side will likely already be frames (either Frame Relay frames or Point-to-Point Protocol (PPP) encapsulated Transport Control

Protocol/Internet Protocol (TCP/IP) packets. Chapter 8 will concentrate on these types of protocols.

Wide Area Networks (WANs) are primarily considered to be circuit-switched networks. In other words, the connection is not permanently in place (route not in place). The PSTN is certainly a WAN, but it is also possible to have secondary access to multiple WANs, particularly if BRI ISDN was passed in the LS0 channel from an ATU-R. Parallel to the analog signaling of "pass-through" signaling with a 56K Modem, the D-channel and B-channel(s) can be passed through the DSLAM to the network-side BRI ISDN port of the network. From there, it may be switched onto the PSTN, Frame Relay network, X.25 router network, or wherever the ISDN has connectivity.

ATM, a Broadband ISDN (B-ISDN) service, may go directly onto an ATM switch, or server. As will be discussed in Chapters 6 and 7, there is the possibility of either permanent virtual circuits or newly switched virtual circuits. The latter situation may make use of a variant of the Q.921/Q.931 signaling protocol discussed before.

The final access point will be "direct" connections to service providers. This is very wide in possibilities although many such services will probably make use of PPP framing to allow for further connections through LANS at their ends. However, direct connections to ISPs may be desired as well as direct connections to video library servers or the equivalent of cable television providers. Chapter 4 will expand on how the various pieces fit together to provide a complete, useful, service to the subscriber.

4 Architectural Components for Implementation

There are many common components to a protocol stack, making it easier to learn a new protocol. It also makes it easier to interwork between protocols and systems. The Open Systems Interconnection (OSI) model is a system of separating the different tasks of a protocol stack into discrete modules. This architecture allows protocols to be designed in advance with the ability to interwork easily.

The hardware, particularly the semiconductor chip sets, is particularly important in a high-speed protocol or access system. These components deal with the interface to the communication network and the handling of physical layer (and sometimes data-link layer) protocols. This hardware must interact with the "outside" world to be useful and there are several main methods of doing this.

Once the protocol architecture has been chosen and appropriate hardware made available in the system, it is necessary to decide just how the system will be used and what other systems must it interact with. Finally, the protocol stack is "connected" into applications so that the main purpose (i.e., Internet access, video file transfers, live audio feeds, data file transfers) can be achieved.

4.1 THE OSI MODEL

The International Organization for Standardization (ISO) is involved with global standards. They are involved in many different areas—from standards for photographic film (they have created the standards for the ISO number on film) to wiring gauges to physical interface components to software architecture.

The OSI model falls into this last category. The main idea behind the "model" is that, if everyone designing protocol stacks uses the same general design principles, it will be easier to standardize and also easier to interwork between protocol stacks. The OSI model is primarily oriented towards communications protocol architectures, although the concept of layers within a system can be used for many different software architectures. Note that the idea of layering is particularly new—the "onion" model has been in use for self-contained systems, such as operating systems, for more than 20 years.

The OSI model is made up of seven "layers" as seen in Figure 4.1. Each layer has specific requirements and responsibilities. Adjacent layers may be "subsumed" into one another when the protocols involved with the layers are not, in themselves, sufficiently complicated to warrant the overhead involved with using interlayer primitives. However, this is an optimizing temptation that interferes with the ability

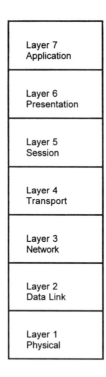

FIGURE 4.1 The Open Systems Interconnection (OSI) model.

of the stack to interact with other protocol modules. Once two layers are combined, there is no effective way to make use of substitute layer protocols.

4.1.1 Layer 1 (Physical Layer)

The physical layer handles the electrical and mechanical requirements necessary for transferring data between two nodes in a network. ITU-T Recommendation I.430 is used for BRI ISDN, I.431 for PRI ISDN, I.432 for STM-1 carried ATM cells, ANSI T1.413 and G.992.1/G.992.2 for ADSL, etc.

The physical layer is also oriented toward general data needs—not just mechanical or electrical. Thus, the physical layer can be involved with optics and lasers to allow for physical manipulation of light for data needs. It can use electromechanical mechanisms such as reading magnetic tape or disk services.

Generally, when the OSI model is applied to telecommunications systems, the physical medium is limited to that which can be used over a distance—electrical, optical, and broadcast media. Often the "primary" recommendation will use the electrical medium as an example, with the translation into other forms being done as a "standard" translation (that is, without specific regard to the protocol) from a general mapping of the characteristics of one system to another.

The physical layer is really two sublayers. In the ADSL documentation, it is split directly into the Physical Medium (PM) and Transmission Convergence (TC)

Architectural Components for Implementation 69

sublayers. The sublayers can be called the physical medium (to use an already documented nomenclature) and the physical protocol. The PM is concerned with the mechanics involved with generating and detecting, specific forms within the physical medium. The physical protocol sublayer is involved with the protocol needed to synchronize the physical layers between the peer entities.

Synchronization of physical layers is often called "activation." This is basically the process of matching up the sampling and generating functions, so that the start of the data stream will be properly recognized. This has two main components: a physical framing requirement which allows detection of the beginning (and end) of a physical frame and information forms that allow interpretation of data within a frame. This will often take two general steps, but it may be "just" a matter of generating a varying form (slowed down or sped up) until it is acknowledged as understood by the receiver.

When the physical layer must interact with a higher layer, there must be a software entity that provides the ability to communicate with the higher layer. This is often called the Low-Layer Driver (LLD). It is also sometimes called a Low-Level Driver or a physical driver. It has the responsibility of handling the primitives passed between the data link (or higher) layer and converting it into the physical level instructions (usually in the form of register settings) needed to carry out the command. This will be discussed in greater detail in Chapter 5.

4.1.2 Layer 2 (Data Link Layer)

The Data Link Layer has the responsibility of handling the protocol needed for error-free communication between adjacent nodes in a network. Note that a data link protocol may supply the appropriate procedures needed for error detection and retransmission but not make use of it. For example, the procedures used in the Frame Relay data link layer for Permanent Virtual Circuits (PVCs) do *not* make use of retransmission.

Generally, it is always possible for higher-layers to take over functions of lower-layers (but not often vice versa). This is a "proper" redesign of the layers based on the specific needs of the system. For example, the data link layer for Frame Relay PVCs does *not* make use of retransmission because it assumes a high level of integrity at the physical layer. If this assumption of the condition of the lower layer is incorrect, then the basis for the design of the data link layer (and other layers) is in error.

A layer 1 entity will hand up data (or other primitives) to the data link layer and basically say "here are some data." The contents and meaning of the data are unknown to the physical layer. Sometimes, some of the simple (but potentially time-consuming) repetitive tasks of the data link layer will be "pushed down" to the lower layer. An example of this is the High-level Data Link Control (HDLC) common frame aspects. The HDLC is a data link format structure. However, the aspects of data link format fall into the "gray" area for physical layer protocol aspects.

The detection of the beginning, and end, of a frame is a responsibility of the physical layer. However, there can be two levels of frame delineation. One is on the physical level, where a certain electrical (or other translated physical medium) pattern indicates the beginning of frame synchronization. The other is a data pattern that marks the beginning of usable data. Within HDLC, this is the detection of a "flag" character defined as the binary pattern '01111110'. Although, in purity, the physical layer is only concerned with whether each datum is a '0' or a '1', it is *much* easier and more effective to have the physical layer look for this pattern while examining the physical patterns for valid binary, and other physical, patterns.

The HDLC format requirements of beginning and ending a frame with a flag means that there must be some type of "escape" mechanism to allow the data '01111110' to be used as data *within* the frame. This is done by what is called "zero-bit insertion." Thus, if a datum of '01111110' is desired to be transmitted, the HDLC transmitter will insert a '0' after five consecutive '1s' *before* putting the flags onto the end of the frame. A pattern of '01111110 011111010 01111110' can thus be detected (and recreated on reception) as a single byte of value '01111110' delimited by flag characters. Without the zero-bit insertion, there would be no way to distinguish between a flag pattern and data of the same format.

The third aspect of HDLC framing often done by the physical layer device is Cyclic Redundancy Check (CRC) generation and checking. In all three of these cases, the duties "pushed down" into the physical layer are involved with groups of physical patterns: flag detection/generation, logical flag escape methods, bit accumulation and calculation. While not specifically associated with the physical layer, they are duties that can be efficiently assumed (and are sufficiently repetitive) by the physical devices used in manipulation of the physical medium.

Similar duties are assumed by the TC sublayer of the physical layer in ATM models. The Header Error Control (HEC) field is very similar to the CRC of HDLC protocols. In this way, we can see that design of protocol stacks can "push down" functions and let upper layers "assume" features if the layers can be relied upon to provide the conditions needed for later assumption. However, this means that the layers are not totally independent. A data link layer using an HDLC protocol must have physical layer devices that directly support HDLC to push down these functions. It will not be possible to use a different physical layer without reassumption of the requirements back into the data link layer.

4.1.3 Layer 3 (Network Layer)

Network control functions are handled by layer 3 of the OSI model. Such functions include call setup and termination, routing, accounting, and higher-layer logical link control in older protocols such as X.25. The Internet Protocol can be considered to be a Network Layer in many models. However, the IP can also take on the data link layer responsibilities if it is carried within a physical layer medium that handles *some* of the data link requirements (such as address generation and detection). IP over Ethernet is an example of this.

A note should be made about the application of the OSI model to older data protocols. Generally, they were not designed with the OSI model in mind and, therefore, have only an approximate mapping. Also, just as certain layer functions can be assumed by higher layers in situations where efficiency and commonality allow this to be done, it is also possible to provide the same type of functionality in more than one layer if the lower layers are unknown in their reliability.

This is true for X.25 where both the data link layer *and* the network layer provide error detection and retransmission. It is also often true for older applications. For example, many of the older file transmission protocols (which might be considered to be layers 4 to 7 of the OSI model) such as XMODEM provide their own error detection and retransmission capabilities. Redundancy is the rule when the functionality of the lower layers are unknown.

The bottom three layers are called chained layers because they are actively used between nodes of a network. Most of the time, the contents of any data packets at layer 3 are unknown. However, the data link layer is often terminated between network nodes. The network layer will often be *used* by nodes in a network, with the network layer messages either passed along (after looking into them for applicable network control information) or recreated with information needed by the next node. An IP router will examine the "network layer" IP protocol for appropriate endpoints, but will pass it along, untouched, if it is not the end node.

4.1.4 LAYER 4 (TRANSPORT LAYER)

Layer 4 is a "host-to-host" layer which is normally examined *only* by the end nodes of a communication link. Transmission Control Protocol (TCP) is a common example of this layer. *Gateways* are also often considered to be transport layer software. A gateway will terminate a layer 1 to layer 3 stack and, via the appropriate primitives, pass contained packet data to another protocol stack (this stack may be the same type of protocol as the original stack or completely different).

4.1.5 UPPER LAYERS

The upper layers of the OSI model are known as the Session, Presentation, and Application layers. They are often combined into an application with very little true layering functions. Sometimes the transport layer is subsumed into this grouping. Because layers 5 through 7 (and often layer 4) are transparent in contents to the network, such blending of layers does not usually cause problems with network protocol architectures.

The advantages of the upper layers lie in the area of host applications; for example, a browser is an application layer piece of software. However, it will make use of the presentation layer to allow use of generic functions (such as encryption for "secure" links) for different windows that it is controlling. Thus, the OSI layering model provides a single common application layer module controlling (potential) multiple types of presentation layers which (per window) control multiple sessions. The sessions will then use the same TCP/IP (perhaps over PPP as a data link layer) protocol stack but vary the contents of the addressing and other control header information to allow for multiple windows and threads to be controlled.

The session layer officially coordinates interprocess activity including synchronization. The presentation layer is concerned with general services that may be used by different applications (potentially at the same time). The application layer is concerned with a *specific* application use of data such as interpretation of HyperText Markup Language (HTML) of File Transmission Protocol (FTP) or remote shareware interactive software, or others.

4.1.6 INTERLAYER PRIMITIVES

Separation of protocols into layers provides specific primitive interaction. The ITU-T has taken advantage of this in the creation of interlayer primitives. The primitives fall into four general types: requests, responses, indications, and confirmations. Requests go from higher layers to lower layers and are answered (from lower layer to higher layer) via confirmation primitives. Note that this only applies to the *same* request—if the request is not being "confirmed," it may be responded to by an indication; for example, a connect request might be responded to with a disconnect indication if the connect request could not be honored. Indications and responses are the same type of interaction except that indications occur from lower layer to higher layer and a response may not be required if the indication is purely informational.

Besides the primitive types, the ITU-T named prefixes for the interlayer primitives (based on the lowest layer involved). PH_ primitives exist for those going to and from the physical layer. DL_ primitives are used for interactions with the data link layer and NL_ primitives are used for interactions with the network layer. The management layer primitives can be brought into the listings by just adding an 'M' to the beginning. Thus, an MPH_ primitive is used between the physical layer and the management entity. Figure 4.2 shows basic primitive interactions between layers.

4.1.7 PROTOCOL MODULARITY

When an OSI layer is "pure" (fulfilling all the requirements of the layer), then other layers may be swapped without having to change the OSI layer module. For example, if a data link control module handles its own HDLC flag stuffing/removal, CRC

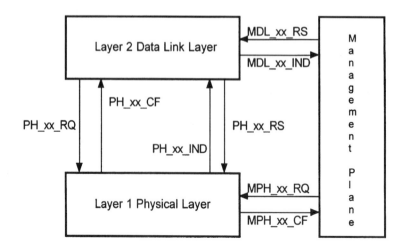

FIGURE 4.2 Example of interlayer primitives.

Architectural Components for Implementation 73

generation and checking and zero-bit insertion, then it will not need an HDLC-specific physical layer. It can make use of any physical layer which provides data transportation across a link (this is sometimes called "transparent" data transmission).

Often a higher layer must know the precise functions and capabilities of the lower layer to be able to control the lower layer appropriately. For example, a network layer requiring multiple logical links must know that this is provided by the data link layer. It must also know how to indicate to the data link layer to set up the logical link and that a particular data packet is to be transported over a specific logical link.

However, the layers should *not* know the purpose of the contents of any data. This "data transparency" is vital for modularity of the stack. Sometimes the "Control plane" can be used as an adjunct to the OSI layer to keep track of just where primitives need to be routed if there is a possibility of multiple layer modules. Control plane functions can be incorporated into a layer module, but should be localized as any use in new stacks will require modification.

4.2 HARDWARE COMPONENTS AND INTERACTIONS

Any physical system must handle two operations: manipulation of the physical environment and control of the manipulation. A hammer has the physical interaction of hitting a nail but must be controlled in order to hit the correct nail. Within a telecommunications system, the interactions with the physical medium will be more subtle than a hammer. However, the basic actions will be combinations of very simple ones. This can be as simple as "raise the voltage," "lower the voltage," "determine the voltage of an incoming signal,"or "change the phase of the secondary signal being transmitted." These are very simple commands but, by combining these commands in different sequences, a huge amount of data can be communicated. Think of the digital waveforms described in Figure 1.4. The physical device is being manipulated very simply. "Set the voltage level at positive 1 for 10 microseconds." "Lower the voltage level to zero." "Maintain the level for 10 microseconds." And so forth. Doing this in the right combination can send complex data, such as a blueprint of a refrigerator, from one location to another.

The actions are simple, but the fact that they must be done so fast means that the commands must be done even faster. Shifting the voltage of an electrical signal in nanosecond time periods is impossible to do via high-layer software. It must be done by semi-autonomous hardware and these are called semiconductor devices based on the technology used. The device may be given a "metacommand" such as "send a pattern of high voltage for 10 nanoseconds, followed by neutral for 10 nanoseconds, followed by low voltage and repeat." The word "repeat" is very important. Fast repetition of simple actions gives electronics (and computers, which use the electronics) their power.

This gets even more complex when conditions are added to the repetitions: "If the incoming wave form is high voltage, send a low voltage at the same period." A combination of simple commands, with the ability to base actions upon conditions and to repeat those commands very quickly, is the heart of every semiconductor device.

4.2.1 Interface Chip

The interface chip is the chip at the heart of telecommunications equipment. The interface chip deals with the physical medium in the manner necessary to perform the physical layer protocol. In the case of ADSL, this means performing the functions of three major functional blocks: Digital Interfaces (DI), Digital Signal Processing (DSP), and Analog Interfaces (AI).

The DI for ADSL will have a DS1 receiver, DS2 receiver, EOC transceiver, downstream multiplexer, downstream buffer, FEC encoder, convolutional interleaver and Trellis encoder for the "downstream" direction. But it will also have a Trellis decoder, upstream buffer, convolutional deinterleaver, FEC decoder, and upstream demultiplexer, in addition to some type of control function to allow higher layers (this time including the LLD in this category) to control the hardware. All of these separate circuits are quite complex, in themselves, but as a coordinated combination we can be glad that there isn't a need to give each circuit a separate command!

The DSP is involved with acting as the "go-between" from the digital interfaces (ASx, LSx, U or S/T ISDN, embedded operational control block) and the basically analog line (or local loop). Depending on the physical code used, this may include splitting into different DMT channels but it will almost always include "constellation encoding and decoding" to put the electrical signal on the line.

The ADSL interface chip is quite complex and, as such, it doesn't provide a good example for this section. A Universal Asynchronous Receiver/Transmitter (UART) chip can provide a much simpler example. UART chips are used in computer systems to provide support for "serial" ports. They typically are attached to ports according to the RS-232C wiring connection. The wiring for such a connection, as indicated by ITU-T Recommendation X.20 bis, is shown in Table 4.1.

The leads for a UART have one lead per direction of data transmission. Ready for sending (or "clear to send") gives a flow control mechanism. Data terminal ready and Data set ready give a simple "handshake" method for synchronization between

TABLE 4.1
X.20 bis Wiring Circuits

Number	Designation
102	Transmitted data
104	Received data
106	Ready for sending
107	Data set ready
108/1	Connect data set to line (used for switched data network service)
108/2	Data terminal ready (used for switched data network service)
109	Data channel received line signal detector
125	Calling indicator (not provided in leased circuit service)
141	Local loop back (not provided in those networks which do not provide automatic activation of the test loops)
142	Test indicator

Source: From ITU-T Recommendation X.20 bis.

Architectural Components for Implementation 75

a computer and serial equipment. The data can travel in a serial fashion over each lead (simplex per lead) and can be any simple electrical binary pattern. The duties of the UART are to basically allow for parity, "start/stop" bits for flow control and different numbers of bits per data unit.

4.2.2 Physical Layer Semiconductors

HDLC devices were mentioned earlier in the chapter. This is an example of a non-interface device that can provide hardware assistance to other layers of software. CRC calculation can be provided by physical layer semiconductors (even if the data link layer maintains official control over such). A physical layer will usually have the interface duties and also the TC type of sublayer. This sublayer may provide HDLC control, UART control, CODEC translation of digital speech form to analog (and vice versa), and other physical translations.

Semiconductors can have different degrees of integration within the part or sets of parts. For example, it is possible to have a single integrated chip which provides a BRI ISDN 'S/T' physical interface, does HDLC framing, CRC generation and detection, provides speaker and CODEC telephone support, and allows for power detection and maintenance on the equipment. It is also possible to find each function on a separate device.

The design choice is primarily a matter of economics. If a highly integrated chip provides five needed functions for less than the cost of separate chips for each function, then it is cost effective. The "real estate" (space needed on a circuit board) may also be important for some applications. For example, an integrated chipset may provide 15 separate "functions" which takes up 1 square inch of space. If only three functions are needed for an application and take only a total of 1/2 square inch then it may be necessary to use the separate devices.

4.2.3 System Configuration Design

The ability to send and receive data across a physical interface is good. To be useful, however, it is necessary that the data be transported for a purpose. The purpose is generally the responsibility of higher layer software — either an application or a "control program." This "intelligence" can reside in one of three places as described in Figure 4.3. It can be on the physical interface circuit board, in which case it is considered to be a standalone system design as no other hardware or interfaces are needed during normal operation. The control processes may be divided between different processors. For example, the transport through the application layers may reside within the data space associated with a personal computer and the protocol for the lower three layers exist on a separate circuit card. The host microprocessor can communicate with the "coprocessor" card via I/O requests, memory mapping, or special data bus commands.

The final general type of configuration is sometimes called the "dumb" board application. Semiconductor devices reside on a circuit board providing the ability to do all physical layer manipulation and protocols. However, all the controlling protocols (not part of the semiconductor device) are located within the host processor. This is a host-controlled system. All three types can be useful in different situations.

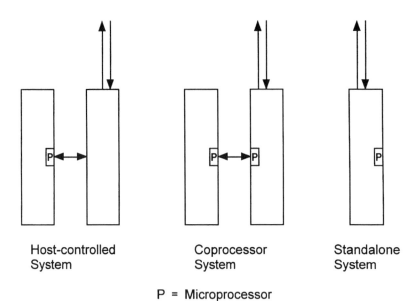

FIGURE 4.3 System configurations.

4.2.3.1 Host-Controlled Systems

A host-controlled system places all controlling algorithms and protocols onto the processor of the host microprocessor system. This type of system is becoming more popular as the speed increases for host computer PC systems (for all types—Macintosh,™ Windows,™ UNIX,™ etc.) and the cost for very powerful computer systems drops.

A host-controlled system is most applicable when a powerful host processor is available and has extra capacity that is not needed for other purposes. A "dumb" card can be very inexpensive. Thus, economic savings are achieved when the multipurpose general computer can be used in conjunction with the relatively inflexible components on the circuit card. It can not be used in situations where the host processor does not have enough processing power to achieve the protocol needs. It is also not a good alternative if the host processor is basically dedicated to serving the interface card. In this case, it will be more practical to buy a "smarter" (but more expensive) interface card so that the general purpose microprocessor can still be used for original application purposes.

4.2.3.2 Coprocessor Systems

A coprocessor system splits the protocol and application needs across two (or more) microprocessors. The interface board, for example, might control the "chained" layers of the physical layer through the network layer and higher layers controlled by the applications on the host system. This is often a good compromise.

Whenever two processors are in use, there must be a control protocol in effect between them. This additional protocol may be very simple or it may be very complex. Generally, the more autonomous the interface board is, the simpler the interface can be.

4.2.3.3 Standalone Systems

A standalone system can be used in telecommunications regions when it has a fixed purpose. A router may be such a device. Once it has been "programmed" with its routing tables and protocol directives it can act by itself. Such a system is only standalone in operation, however. There must be some type of interface that allows control and administration. Many network administration programs have built-in protocols to allow remote control of routers and other network devices.

Most standalone systems have a direct system of I/O. A good example would be a television receiver, which can be programmed via an infrared controller but, when not being programmed, act autonomously according to prior directions. Another example would be a traffic light control system. It will have specific requirements which may include information that is provided by specialized sensors in the road. It may also have changeable state event tables based on time of day (different patterns for rush hour or for school days). This will be discussed more in Chapter 5.

4.3 PROTOCOL STACK CONSIDERATIONS

The first protocol stack consideration is what protocol (or protocols) are desired. The next is where the protocol is controlled. The third is how the coordination of protocols is to be accomplished. This first category includes whether the protocol is to be used only with a fixed destination. If multiple destinations are to be connected to, it is necessary to have signaling protocols and methods to control them. If the system is intended to interact with other networks (either an intranet/internet situation or a routing situation) then interworking protocols will be needed. Normally, interworking protocols will be contained within the network node equipment. Finally, if multiple protocols are to be used within the same configuration, a method of setting up the appropriate stacks is required.

4.3.1 SIGNALING

Generally, the host microprocessor will control the final destination. This is similar to a user dialing the digits on the phone to make an analog speech call. In the case of an Internet browser, the application will check to see if a connection has already been established. If not, it will initiate the connection to an Internet Service Provider (ISP) using either default (pre-administered) information or will prompt the user for destination number and any protocol permission information (such as user identification and password).

Once the user, or application process, has initiated a connection request, the request will pass down the OSI layers until it can be accomplished or an indication comes back that indicates the request is impossible. A situation that would cause the latter would be if the phone line was already active ("busy").

The application layer initiates the connection which is passed down to the network layer which tries to initiate the data link layer. The data link layer sees if the physical layer is active and, if not, will request activation of the physical interface. Once physical line activation has been achieved, the data link layer will negotiate its protocol, then notify the network layer that it has completed the request. The network layer then proceeds to do anything that it needs to do.

This type of downward progression followed by upward confirmations is typical of layered protocol systems. Each lower layer must be ready before the higher layer can do its function. Thus, some type of activation or service request, or indication, is needed between the layers to fulfill the original request.

4.3.2 Interworking

Interworking is the process of having two (or more networks) interact. Generally, if interworking is to take place, some conversion of protocols or data is necessary. Thus, a router connecting two LANs which are both Ethernet networks is a very minor form of interworking (which, for some, might not even be considered an interworking situation). At the most, all that will be needed is a change of address form or some parameters within the protocol (for example, if each network uses a different maximum frame size). The protocol, itself, remains the same.

Interworking might occur between an analog and a digital network. In this case, both signaling and data must be converted. The dialed digits would be converted into the signaling protocol used by the digital network. The analog speech patterns would be converted into a digital form such as Pulse Code Modulation (PCM) which is created by sampling the analog wave (8,000 times per second to cover the data at a rate of twice the greatest potential change).

Another form of interworking might involve conversion of signaling, data, and protocol information. For example, a frame relay network might interwork with an X.25 network. When such disparate networks connect, there will be some direct mapping between the protocols. There will also be aspects of the protocol which do not exist in the other protocol. In this case, it will be necessary to "terminate" that portion of the protocol (perhaps with notification back to the originator) and acting according to "default" (predefined as what is done if nothing else is specified) parameter sets.

4.3.3 Stack Combinations

Interworking may also involve multiple protocol stacks. An example of this might be a Basic Rate Interface Integrated Services Digital Network (BRI ISDN) frame carried within an ATM/ADSL network. This is a situation of *encapsulation*, which is a special form of interworking. This is similar to routing a letter around a corporate department and, at the end of the routing, putting it into an envelope to mail to another corporate location. Once it has arrived, it will be removed from the envelope and routed once again.

Architectural Components for Implementation 79

Intranet use is increasing within the corporate environment. However, for larger corporations and for telecommuting corporations (which are increasing in both number and percentage), the physical location of the employees means that not all employees can be on the same LAN at the same time. There is a need to connect the corporate LANS together but still act as if they are on the same network. This type of encapsulation is called "tunnelling" because one network protocol is encapsulated within another network protocol.

Tunnelling is primarily needed to preserve security features within the corporate intranet. Thus each node on the intranet—whether remote or local—are provided with the same security and application features without direct knowledge or interaction with the WAN Internet.

4.4 APPLICATION ACCESS

The application must have access to the hardware and protocol stacks, whether it is a host-controlled, coprocessor, or standalone system. We briefly mentioned I/O request access and memory mapping. For high-speed data transfers, these methods may not be adequate—I/O requests being too limited in data space and memory mapping being too expensive for large areas of data. (A memory mapping allows two processors to access the same area of RAM or ROM with different physical addresses for the memory.)

When the host needs to access the data, there will be some type of physical data interface between the host and the interface board. Since this is over very short distances (possibly inches or centimeters) there is little need for data link layer types of protocols. There will be need of "flow-control" and buffering. Flow-control keeps the two (probably different speed) devices in sync.

4.4.1 HOST ACCESS

Host access will be discussed in greater detail in Chapter 9. The main approach is to use some type of network protocol or data bus design so the host can treat the data as local. Data buses carry data between devices on a computer system—with local "bus stops" at each device. This is a general-purpose type of memory mapping where there is normally only a single mapping of address space. For example, the region between hexadecimal 0x00000000 and 0xDFFFFFFF might be located within the RAM on the main processor board. The region from 0xE0000000 to 0xE4FFFFFF might be on a different board controlling video applications and so forth.

4.4.2 CONTROL SYSTEMS

A control system is an autonomous system in normal operation. Real-time interactions may be needed however. A non-telecommunications example would be a train routing system. If a switch is stuck in the wrong position, it is important to know immediately. However, in general only polled queries are needed. No large amounts of data are moved.

In this situation, there is no need for large data movements; only for speedy transfer of small data blocks. This is a special type of standalone system. The older types of I/O may be sufficient for this. In Chapter 9, we will examine how the hardware is accessed and controlled in greater detail.

5 Hardware Access and Interactions

When high-speed access is discussed, it is almost always synonymous with semiconductor chip usage. The hardware will probably have a number of components, many of which are semiconductor devices; for telecommunications needs, there will be a physical interface chip. Other possible components are Random Access Memory (RAM), Read-Only Memory (ROM) or, more likely, Erasable Programmable Read-Only Memory (EPROM) or "flash" memory. Other devices on the circuit board will include "glue logic" which allow the various signals of the different chips to be synchronized to act as a single system and which adapt clock leads and shift voltages and currents for the different requirements of the chips.

Most existing "narrowband" protocols do not require hardware assistance to perform the necessary work above the physical layer—although HDLC protocols may very well take advantage of the capabilities of semiconductor chips. When we get to protocols such as ATM, however, even when it is operating over "slow" ADSL, hardware assistance may be imperative.

Most interfaces into ATM supportive chip sets will probably interact at the Convergence Sublayer (CS) part of the ATM Adaptation Layer (AAL). Although there will be commands to direct both the ATM and physical layers, the speeds are such that direct control of the ATM and Physical layer responsibilities will slow the process too much. The Segmentation and Reassembly (SAR) function of the AAL is rather parallel to HDLC hardware support for the HDLC narrowband services. It is something that the general purpose microprocessor probably *can* do but it is something that hardware can do more easily without degrading performance of the microprocessor.

The final set of devices that will be needed on a circuit board (whether it is a "plug-in" card or a separate box with its own power supply) is support for communication with a host or user. This is not only needed for situations where the system is fully standalone (and it is likely that it will be desirable to be able to administer even routers remotely). Figure 5.1 has a "generic interface board" architecture.

Although the various devices on a circuit board will have different functions, they are accessed in similar ways. This chapter will mainly discuss how the hardware will be accessed by software and the types of algorithms needed to properly interact with the semiconductor chips.

It should be mentioned that, when put together, the semiconductor devices will "talk" to each other without using the procedures and methods discussed in this chapter. They will interact by direct use of "pin-outs." These are circuit wires which

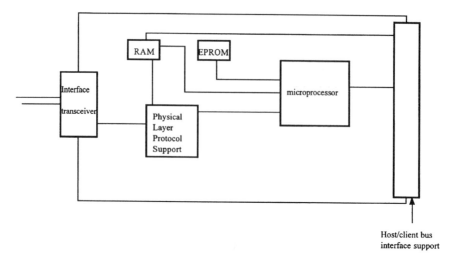

FIGURE 5.1 Generic interface board architecture.

connect to the "outside world" (not within the circuitry of the chip). These pins will be connected to the pins of other chips (perhaps with "glue logic" making them compatible). The information passed along in this way is basically a "high" or "low" electrical level—but this can be interpreted as "on" or "off" and much information can be passed successfully between the semiconductor devices.

5.1 SEMICONDUCTOR ACCESS

Semiconductor devices are accessed via the pin-outs connected into the circuitry which can be connected to other circuits. Many of these pin-outs are concerned with power, clock rates, and other timing circuits to coordinate the actions of the chip with other devices.

The pin-outs will often be grouped into function groups. These include general purpose leads, timers, interrupts, clocks, bus control, bus arbitration, system control, interrupt control, chip select, and power groups. General purpose leads can usually be configured for a specific use or for "general" use for software to be able to use the pins for control of other devices (to which the electrical leads are connected).

Timer leads are available on microprocessors which have an internal clock state. These allow signals to be sent at specific intervals (which can be programmed). A common use for a hardware timer is to allow polling on a periodic basis. Another purpose is for "watchdog" use which allows the hardware to override whatever the software may be doing (for example, the software may be "stuck" because of logic errors), examine the condition of the system and, if needed, start corrective actions.

It is possible to use the hardware timers to control protocol events. For example, the data link layer transmits an information frame and expects an acknowledgement within 3 seconds. It is possible to have the timer expire in 3 seconds and see if the acknowledgement has occurred. It is more likely that the timer will be "reset" or turned off if acknowledgement happens. Thus, the timer expiration will be an indication that acknowledgement was *not* received in the appropriate time frame.

Normally, there will only be one or two hardware timers available in a general purpose microprocessor. With such limited resources, it is not feasible to make use of them directly for protocol needs (since it is likely that more timers will be needed than the timer hardware can provide). In this case, the hardware timer can activate a *software timer*. As an example, the hardware is set to time-out on a periodic basis (restarts once it has expired) every 10 msec. This can cause the activation of a software function that, in turn, inspects a list of timer events. If any have expired, it can then send software messages to notify the appropriate software that the timer has expired. This method can extend the timer architecture to allow for many timers.

Interrupt control groups and interrupt groups are both concerned with *events*. The events are items for which the device has been designed to detect. This might be the end of an incoming frame or the end of an outgoing frame being transmitted. It might be an indication that the physical line condition has changed. Many times, interrupts can be enabled or disabled. An interrupt has a high amount of overhead because it must cause the existing context (the contents of all registers for the microprocessor) to be saved so that, when the interrupt has been processed, it can continue with the work that was being done before receipt of the interrupt. Therefore, it may be more efficient to *poll* for events. This is basically "looking" at whatever information is provided to see if an event has occurred.

The use of interrupts and polling is a tradeoff. Interrupts require more overhead *when the event occurs,* but do not take any active maintenance when nothing is happening. Polling events take much less overhead, but must be done on a periodic basis (the length of which must be determined by the characteristics of the event) so that the *total* amount of time spent may be greater than that used within an interrupt routine.

The decision to use an interrupt or polling structure ends up being a mathematical comparison. The important parts of the formula are: average time between events, length of time in which the event must be handled, amount of overhead for an interrupt, and amount of time needed to poll the event. The total amount of overhead per method is the overhead times the number of events per unit period; e.g., if an event occurs 3 times a minute, and overhead for interrupts is 50 msec, then the total amount of overhead using interrupts is an average of 150 msec every minute. If the event must be handled (serviced) within 2 seconds, then a poll must be done once every second (to make sure that 2 seconds is not exceeded). If it takes 10 msec to check the event then a polling method will take 60 times 10 msec or a total of 600 msec overhead every minute.

Bus control and arbitration leads are concerned with ensuring that information can be communicated between devices and that the information doesn't come from two (or more) sources simultaneously (which would corrupt the data). Chip selects work with the address bus to determine where the information will go. Power groups are concerned with supplying the circuitry with the proper electrical needs to function properly. Clock leads synchronize events within, and between, devices. Finally, system control can be used to restart the device to a known state, or halt the processor, or indicate an illegal use of the instructions.

Access of the control and data mechanisms of a semiconductor chip requires a sequence of events to occur. The sequence is usually something like the following. A lead is "seized" (brought to a high or low "active" state). At this time, no other device may seize the lead. Control information is passed to prepare the device for the data—this may be a register number or a memory location or a command. Once the device has been seized and prepared for the data arrival, the data are sent. The data may be acknowledged or unacknowledged. After the sequence is completed, the seized lead is released, allowing other devices to access the chip. This sequence is normally done as a consequence of some other instruction. Common methods for doing this are via memory maps or I/O requests.

5.1.1 Memory Maps

An address space is the range of identifiers that may be accessed by a device (i.e., from location 0x0000 to 0x3FFF). A memory map works by dividing the address space into separate segments. Each segment may be further broken down into smaller ranges. However, the memory map works to allow devices to access other devices—the subranges are usually only used *by* the addressed device.

As an example: a microprocessor needs to access information within an interface chip. The memory map indicates that all I/O performed between 0x4000 to 0x43FF (a range of 1024 identifiers) will be routed between the requesting device and the interface chip. The "address decoding logic" indicates that when this address is written onto the address bus, a sequence of instructions will commence. This will probably be something similar to the above mentioned sequence. The net effect will be that a write of data to address 0x4204 will go to the interface chip and that, because of the specific location within the address range, some type of behavior is expected as a result of the write.

5.1.2 I/O Requests

An I/O request is very similar to that of a memory map. The difference is that the I/O designation is unique for the system or for the circuit board. For example: a personal computer may have a data bus with different circuit boards connected via a data bus control plane. Each board could have its own I/O interrupt request line. A machine instruction (such as inpw) will generate a request for a particular I/O request device and then read the data associated with it.

It's possible to do the same thing on a circuit board—allocating each device its own I/O interrupt number (or numbers, since it can control all devices on the board and can, thus, allocate as desired). Therefore, a command from one device to another becomes an I/O request followed by data. The data may include address information.

5.1.3 Registers

Registers are the normal grouping within the command structure for a device. Registers may be internal—memory areas used for internal calculations and purposes and not directly accessible by any other devices. They may also be "hardware/software interface" registers. These registers may be read-only, giving status of events

Hardware Access and Interactions 85

and command conditions. They may be write-only (being only a virtual location for access) or they may be read/write such that data written can be read back from the same location.

Command registers will usually be split into "bit fields." That is, each bit of the register will indicate a certain action or event; e.g., there may be a byte register with five fields. One bit indicates whether the clock lead is being used as input or output (internal clock generating the signal or being given to the device by some other semiconductor device). Another four bits might be a "clock divider" which allows the clock lead to be adjusted for disparate device needs. Another bit might indicate that the clock is to be used for a timer. And so forth. The register will always occupy a "logical location" but may not exist as a physical entity. Say that writing the top bit of a control register will reset the device. Is there any reason to have a real location for the bit to be stored? It is much easier to have the address decoding logic cause the device to be reset than to store an intermediary piece of data.

5.1.4 Indirect Register Access

Indirect register access occurs in two steps. The first is a command setting up the register address for use. The second is executing the command followed by data (if any). This is similar to having a central switchboard that must be notified of the need to communicate to a particular endpoint before talking on the phone. This method simplifies hardware because the routing algorithm can be simpler (since it is directly controlled by external events rather than as a consequence of interpreting external events).

5.1.5 Data Movement

One of the important tasks that can be given to a semiconductor device is data movement. This doesn't apply to all devices. Often the interface device will take the data input lead and put it into the form needed by the physical interface. If data are not present on the lead, it will insert whatever is suitable for "idle" units of data. However, an AAL/ATM device or an HDLC device will expect data to be passed directly between the layers.

There are two main methods of transmitting and receiving data. One is to use a First-In; First-Out (FIFO) queue that is usually located as a logical set of registers accessible by the semiconductor chip and the software control. The other is the use of buffer descriptor lists.

5.1.5.1 FIFOs

A FIFO will have a specific size. Although this size may be settable by software, it will have the same size until a new command is entered. Often, the maximum size will be limited by the hardware design. There will be three events associated with the transmission FIFO. These are "FIFO empty," "FIFO below threshold," and "FIFO full." The FIFO full event may need to be kept track of by the software; e.g., the FIFO is 20 bytes long. When the FIFO empty event occurs, the software will know that twenty bytes may be inserted "into" the FIFO.

The "FIFO threshold" event is important to make sure that data are always in the FIFO. If the FIFO is depleted before the complete frame is sent out, there will be idle cells or characters inserted into the midst of the frame—which will usually be considered an error by the other end. This is called an "underrun" condition. Setting the FIFO threshold to a level above empty gives the software time to insert more data so that it is not depleted before the frame is complete. The end of the data is usually marked by the software setting a special bit in the transmit control register (saying that the last of the data is in the FIFO).

A received FIFO may have the same three events associated with it. The critical one for reception is "FIFO full." If at least one datum is not "taken out" of the receive FIFO between the time that FIFO full is set and the next byte (or data unit) is received, an "overrun" condition will occur. Overrun indicates that data were lost because there was no place to put the data. Events can be polled or, often, they can trigger interrupts. An example of FIFO usage, and short algorithms, can be found in Figure 5.2.

5.1.5.2 Buffer Descriptors

A buffer is an area of memory where a group of data can be located. The buffer may be in an area of RAM directly controlled by the device or it may be in general-purpose RAM for the circuit board.

Buffer descriptors are more commonly used by devices which are controlling high-speed protocols and interfaces. A buffer descriptor allows a set of data of

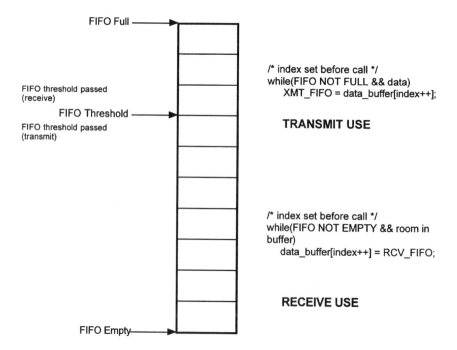

FIGURE 5.2 Example of a FIFO data queue.

Hardware Access and Interactions

indefinite size to be associated with the receive or transmit process. Buffer descriptors will often be linked together, so that when one buffer is used, the next available is immediately usable. Thus, the only way that an "underrun" or "overrun" condition can occur is if all possible buffer descriptors (and associated buffers) have been used.

The buffer descriptor will usually have two parts. One part will describe what is to be done with the data (whether it is the first part of a string of data, for example, or what transmission mode is applicable). The other will be the address (pointer) of a buffer area and the length of the valid data at that location. The third (optional) part is a pointer to the next buffer descriptor or an indication that the buffer descriptor pool should be restarted. Figure 5.3 gives an example of a buffer descriptor table.

5.2 LOW-LEVEL DRIVERS

A Low-Level Driver (LLD) acts as the software interface between a semiconductor device and the controlling software. If it is used within a data protocol stack (using OSI layers), it will usually act as the boundary between one layer and another. The LLD must be able to interpret primitives from, and send primitives to, the software module (or modules) with which it is associated. It will also be involved with the

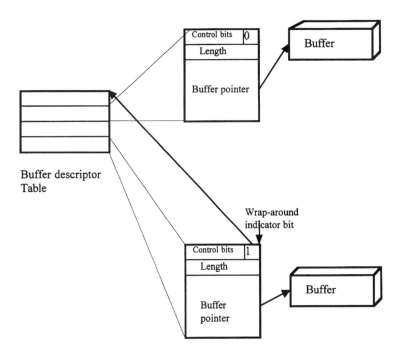

Buffer descriptors

FIGURE 5.3 Buffer descriptors and data transmission.

handling of events that occur associated with the hardware. Handling of events will either be part of a polling (keep checking on a periodic basis) loop (perhaps triggered by a hardware or software timer) or part of the LLD will act as an Interrupt Service Routine (ISR).

Thus, the LLD will usually have three parts. The first part is initialization. This must occur before the LLD is used (or interrupts, if used, enabled). Initialization will set up the control registers of the hardware device based upon how it will be used in the specific application. This is often done as part of a "power-up" sequence. The second part will handle primitives coming from the software module (or modules). The third part will handle hardware events.

5.2.1 Primitive Interfaces

The primitive interface will usually have two parts. If it is part of an OSI layer stack, there will be standard defined primitives that must be used. There will also usually be commands that are special because of the specific purpose of the hardware.

5.2.2 Interrupt Servicing and Command Handling

When a command from a higher layer is interpreted, it will usually involve the manipulation of registers in the hardware device. During this period, "race conditions" can develop. A race condition occurs when two or more entities want to access the same data (or register) at the same time. There must be some way to synchronize the desired actions of the entities so that a complete (and logical) sequence is completed before the other entity starts to manipulate the data. If one entity is interrupted before it has finished, data structures may be left in an incomplete (and error-prone) state.

So, in addition to interrupts, microprocessors and other devices offer the concept of an "interrupt mask." An interrupt mask tells the device what interrupts are permissible. This may be set up at the end of initialization. When a command is being processed, and it is dealing with the hardware registers, the interrupts should be masked (or all interrupts disabled) during this period to prevent a race condition.

This period, when interrupts are disabled, must be as short as possible, otherwise, there is danger of overruns, underruns, and other error conditions. Normally, if an ISR is used, further interrupts by the same device will be disabled. This means that the processing during an ISR must also be short. Since the actions performed by a polling loop and an ISR are basically the same (but with different overhead requirements), there is no inherent advantage in one method of event servicing over the other; however, in both cases, the register accesses must be protected. Use of asynchronous message passing may help in keeping the interrupt disabling time short.

5.2.3 Synchronous and Asynchronous Messages

A synchronous message is acted upon before the calling routine continues with other tasks. This is like ringing a bell and waiting for the door to be answered before handing a letter to a person. An asynchronous message begins an event, but doesn't

Hardware Access and Interactions   89

wait until it finishes before doing other things. This is similar to posting a letter. The letter may take an indefinite period of time to arrive and, upon arriving, may go noticed or read only when the recipient is ready to do so.

Operating systems, or tasking systems, will usually provide support for asynchronous messages. The sender requests that a message be left for a task. This is then marked in the status associated with the task and, next time that the recipient wants to check to see if there are messages, the message can be relayed.

Synchronous messages are good for software layer to hardware layer interactions. Time must be minimized anyway (for the reasons just discussed) and there will be less overhead with a synchronous message. Asynchronous messages are often appropriate for hardware to software data unless the time needed to react to the information is known to be very short. This keeps the hardware software (which is probably interrupt protected) as short and quick as possible.

5.3 STATE MACHINES

The core of most protocols is a "state machine." A state machine is a rigidly defined set of rules that defines just what action is to be taken when a certain event occurs. This combination occurs from use of a closed set of states, events, and actions. An event which is not defined within the state machine either cannot occur or it must be ignored if it does occur. (Of course, designers of state machines do make mistakes and state machines can be modified in accordance with new information.)

Protocol state machines are applicable at every level of a telecommunications protocol stack—particularly the "chained layers." A well-designed implementation of a state machine can make a huge difference in the efficiency of the stack.

5.3.1 STATES

A state is basically a history. Naturally, software doesn't really have a history in the same sense people do; however, it is possible to keep track of the events that have occurred and that is the definition of a state. A state is the current condition of a protocol stack, or piece of hardware, based on prior relevant events. (Remember that events which occur, and are not defined as part of a state machine, are ignored.)

A history requires a beginning (whether the beginning is artificial or not). Initialization of an LLD will set appropriate registers, and internal variables, and this initial condition will be considered to be the beginning, or idle, or "null" state. Events which occur *before* being set to the idle state become part of the definition of that state.

Once a state is entered, it will remain the same until some event triggers an action which includes a change of the state. Of course, it is impossible for nothing to occur so, in theory, states should be changing all of the time. That would be true of people because we are unable to limit the number of events that will be relevant to our actions. We must breathe. We must eat. There are many events and actions that can not be ignored in humans. However, for machines, it is possible to remain in the same state for an indefinite period of time if no *relevant* event occurs. A light bulb that is off will stay off forever unless a specific event happens (the electricity is turned on). (It is also possible for some other event to happen so that it is never able to be turned on—such as being removed from the fixture, broken, and so forth.)

In hardware or software, history is recorded as the contents of registers or memory locations. The contents will not change unless an event cause them to change. Change of contents of memory does not, in itself, indicate a change of state. It is possible for events to happen and the state not change, even if it is relevant, as long as the behavior does not change. That is, if a state machine before event A still reacts to events in the same way afterwards, the state has not changed.

5.3.2 Events

Within a state, different relevant events can occur. In a protocol state machine, these are primitives. Note that a primitive may include resource indications and responses such as timers.

A protocol stack is in a stable state. An event happens. What is the procedure that will be followed at this time? First, the event must be identified. This is associated with its *entry point*. For example, a message that is received within the message box associated with layer three to layer two primitives should be a message from layer three (DL_ request or response). The context of the entry point will limit the number of possible events. (If a message from the LLD arrived in the layer three to layer two message box, then it would either be ignored or misinterpreted.)

Each entry point will have a set of primitives which can occur. The primitive format will enable the exact primitive to be recognized. Based on the exact primitive, there will be parameters associated with it. For example, a DL_DATA_RQ primitive will have a link identification, length and data address. It *may* have other characteristics but those are the required parameters. However, the primitive implies other requirements. A DL_DATA_RQ primitive is an acknowledged data transfer. This means that the window must be sufficiently clear to allow transmission. It means that the buffer must be retained, for possible retransmission, until after the acknowledgement is received. It specifies the form of the Layer 2 control field. The link identification is directly mapped into the address field. A primitive must contain enough information for the appropriate action to be invoked.

An event that occurs from the lower layer may require parsing of the contents. A PH_DATA_IND primitive will include data information. This data is checked to make sure that it is a legal data link layer frame. If it is legal, the address and control field information can be used to check the appropriate internal variables to determine the state for the particular link. If it is illegal, the currently existing state will determine the results of the message. If it cannot be identified (because it is illegally formatted, for example) then it must be responded to as if the link was in the original idle state because lack of identification requires a base response.

Timers are a special type of primitive. This is because they are initiated within a module rather than being created by some other layer, or module. Timers are started because an event is *expected* within an interval of time and there are actions to be done if the expected event does not occur.

This is an example of an error condition. We said earlier that an event that is not defined as relevant within a state machine could be ignored. It may also be

Hardware Access and Interactions

registered as an erroneous event. (This can be useful in decisions to revise state tables that turn out to be incomplete.) Error conditions can either be purely informational or they can trigger attempted recoveries. Undefined events, however, can only be logged as informational since, being undefined, there is not way to determine the appropriate recovery method.

5.3.3 Actions

Actions are the product of a state and an event. A particular action is done because of being in a specific state when a relevant event occurs. Actions may include generating new primitives, setting timers, changing internal variables, and changing states. It is rare that change of state is the only thing that is done because there must be conditions that are evaluated that cause a different behavior. Change in conditions implies a change in internal variables.

Sometimes, a state will have *substates*. A substate is a set of conditions within a general state that cause different actions to be done. The difference between a substate and a state is that the action rules remain the same—only the conditions that are involved with the rules that have changed. For example, a data link layer can be in an active state—able to send and receive acknowledged data. If a DL_DATA_RQ primitive arrives and the window is full, the data must be queued. This is an action in response to conditions and a rule. The rule would be something like "if a DL_DATA_RQ primitive occurs and the window is full, store the primitive." Later on, an acknowledgement occurs which allows the DL_DATA_RQ primitive to be processed. Thus, the rule becomes "on receipt of acknowledgment which changes the window, check for storage of DL_DATA_RQ primitives and send if still possible." If the window is *not* full, the rule still applies ("if a DL_DATA_RQ primitive occurs and the window is full, store the primitive") but the conditions have changed such that the window is *not* full.

5.3.4 State Machine Specifications

State machines are usually predefined for the implementor. Of course, someone (or group of people) originally defined the state machine. Creation of a state machine involves starting from an initial state and deciding what state is desired. Intermediary states are defined and the possible events are listed. Next, the process of deciding upon the actions that are invoked upon events in particular states is created.

One form of stating a state machine specification is called a state table. A state table can be thought of as a matrix. On one side (call it the row side) there will be states and on the other side (call it the column side) there will be events. In the intersection of these rows and columns will be actions. Actions will be based upon the intersection (state and event) *plus* the conditions of any variables that have been stored. The state table matrix is a good format to make sure that all combinations of states and events are addressed. If a person is designing a state table, it is also good to have a list of all internal variables also. This allows the list of variables to be examined to see if they provide special conditions upon which actions should be defined.

Another form of a state machine specification is called the Specification Definition Language (SDL). This looks more like a flow chart. Each state is usually represented by a circle. Transitions between states are indicated by arrows going from one state to another. Events that occur which cause the transitions are listed on top of the arrow (the arrow may point back to the same state). Actions are listed at the "bottom" of the transition arrow. This form has the advantage of being able to easily combine semantically similar actions. For example, if event A in state B causes the same actions as event G in state B, it is easy to combine events A and G in an SDL. Both matrix points must be defined in a state table matrix. SDLs are more difficult to check for completion.

Prose can also be used to describe a state machine. This is of the nature "if event A happens in state B then action 3G should be done." Prose is the most versatile mechanism and the least precise. It is easy to say "if event A happens in *any* state, do action 2F." In a state matrix table, it would be a duplication throughout the entire column or row; in an SDL, it would cause a duplication of graphics or an addendum/marking at the edge of the SDL chart. However, it is even more difficult to ensure completion than SDLs. Prose methods are very good at exception cases ("for all states except for state B, event G should cause X").

5.3.5 METHODS OF IMPLEMENTATION

There are three primary methods of implementing state tables: via a direct translation of a state matrix table, a "push-down" automaton, and a hardware implementation. Actually, a hardware implementation is probably a circuit-level translation of the state matrix table because state matrix tables and Venn diagrams are very similar.

A direct translation has the primary choice of being a "row-first" or a "column-first" translation. Assuming that rows are events and columns are states, a column-first implementation will have function entries for each state. Within the state-specific function, there will be decisions made (possibly within a 'C' style "switch" statement) based on events. The other possibility is an event-specific function, with a check for appropriate state. The "best" direction will depend on the specific state table matrix. Examination of the table will see if (perhaps with sorting of rows or columns) any areas can obviously be combined.

For example, if you find that most of the action boxes for a given event are the same for most states then an event-first implementation is probably best because it will have the fewest conditions ("if event A then if state B or if state C, ..."). On the other hand, if there are few common action boxes then implementing a state-first implementation is probably better as there are likely to be fewer states than there are events.

5.3.6 EXAMPLE OF A SIMPLE STATE MACHINE

Table 5.1 shows a simple state machine matrix. This is for a (simplified) traffic light control system. The environment is a simple cross intersection of roads A and B. There are four states indicated as A_DAY, B_DAY, A_NIGHT, and B_NIGHT. State A_DAY indicates that road A has the "green light" and it is during the daytime

TABLE 5.1
A Simple State Machine Table Matrix

	Car sensed on A	Car sensed on B	Car not sensed on B	Tb expired	Tt expired
A_DAY	/	Start Tb	Stop Tb	Cycle lights Restart Tb timer Change to B_DAY	Change to A_NIGHT Restart Tt
B_DAY	/	/	/	Cycle lights Change to A_DAY	Change to B_NIGHT Restart Tt
A_NIGHT	/	Start Tb	Stop Tb	Cycle lights Restart Tb Change to B_NIGHT	Change to A_DAY Restart Tt
B_NIGHT	Cycle lights Change to A_NIGHT	Restart Tb	/	Cycle lights Change to A_NIGHT	Change to A_DAY Restart Tt

period. B_DAY is, once again, during the daylight period but road B has the "green light". A_NIGHT and B_NIGHT are similar but in the nighttime period.

There are five events that can happen. A car is sensed on the road A sensor, a car is sensed on the road B sensor, a car is not sensed on road B, or one of two timers expire. The timers are Tb and Tt. Tb is activated because of road B events. The Tt timer is a continuous timer, which oscillates the current state set between the DAY states and the NIGHT states. No timer is needed for A road events because they always have precedence.

Assume that the starting state is A_DAY with Tt set to expire at the end of the daytime period. Daytime traffic algorithms give high priority to road A (road B is a relatively minor cross street and road A is a commuter street). Sensors on road A do not matter as long as road A has the green light. However, the sensors on cross street B are important. A car being sensed on road B at the intersection will be an event. (Note that one simplification of the table is that, normally, a car must be sensed for a period of time before triggering an actual event—thus there may be *two* timers, Tb and Tpb, for Timer pre-B event.) This event will activate the Tb timer. If the car leaves the sensor before Tb expires, the timer is stopped. (Note that this can happen if a car edges too far into the intersection in anticipation of the light changing!)

If timer Tb expires, an action and a state change occur. The actions are to cycle the lights (cause road A to have a yellow light, then red, then change road B's light to green) and restart the Tb timer (perhaps to a minute). The state change is to B_DAY. This state indicates that it is daytime and that road B has the green light.

When timer Tb expires in state B_DAY, the action is to cycle the lights back to a green light for road A and to change state to A_DAY. Note that Ta and the traffic sensors for road A are never checked. These events are irrelevant for the DAY states because road A has priority. Road B events may cause a temporary change of state but road A will have the majority control.

The Tt timer is involved with shifting between the daytime and nighttime algorithms. In the daytime, the algorithm gives precedence to road A because much more traffic is expected. At night, traffic is much lighter in both directions. The amount of traffic on road B, however, can be expected to be greater *in proportion* to the traffic on road A than in the daytime. This indicates that a different algorithm is needed. It is frustrating (and frustrated drivers can be dangerous) to be stopped at a red light with no traffic going on the other road. Thus, an algorithm, which gives greater precedence to road B, is warranted.

Road A still has greater priority than road B (it is, after all, still considered the commuter road). However, as long as road B has traffic on it and road A does *not* have traffic, the light will stay green for road B. This can be seen by the way that events continue to cycle to maintain a state as long as traffic continues and the other road does not have a significant event.

In this simple state machine example, we have states, actions, and events. This was designed by looking at the environment. What are the main states? One road has a green light while the other has a red light. What events will cause the state to change? Cars arriving at the intersection? Are the actions permanent? Is there any other event which may cause the actions to be different based on the event?

The questions of what and how are the primary ones involved with the design of a state table. (The question "why" may enter into the problem of getting financiers and customers involved.) In a state table, "when" becomes an event based on a timer. "Where" *may* be involved but only based on a sensor.

Actually, the situations involved with when and where are not any different for human activity. Our sensors of our eyes and ears and fingers and nose (probably not taste) are all involved with telling us just where something is located. Thus "where" is involved with what our sensors indicate. "When" is based on internal or external timers and the time precision can vary depending on culture and circumstance.

5.4 ADSL CHIPSET INTERFACE EXAMPLE

Our discussion on LLDs indicated that the ADSL chipset interface was needed to create an interface between the software (running on a controlling general-purpose microprocessor) and the hardware. The hardware interactions are based on access to registers for control or I/O. This means that each LLD is specific to a particular hardware design. The interface, or Application Programming Interface (API), may be designed as more flexible, however. We mentioned that the PH_ primitives that are documented by specification groups can be a consistent core for any LLD that interacts with a data link layer type of software module. However, other primitives will be specific to the hardware.

It is possible for the LLD to be split into two parts: one involved with communicating with the "upper layer" software module" and to more general hardware-related functions and the second providing a mapping of general hardware-related functions to specific hardware registers and interactions. This type of "device driver"

Hardware Access and Interactions

library can make the task of the writer of the LLD much simpler by isolating the specific hardware design from the general functions implemented by the hardware.

An example of this type of device driver library exists for the Motorola CopperGold™ (MC145650) ADSL transceiver. This library supports an API for functions based on ADSL needs (primarily ANSI T1.413, but general enough in design to be applicable to G.992.1 or G.992.2 standards). Note that this isolation from the hardware allows changes to be made without necessarily having to make changes to the LLD.

The CopperGold™ API library splits the functions into five sets. These are involved with initialization, statistics, Embedded Operations Control commands (for ATU-C implementations), general data retrieval including ADSL Overhead Control, and vendor specific "CG" commands. Note that the specifications for ADSL (and ADSL "lite") both allow for specific operations to be vendor specific. These can be used in conjunction with the EOC READ (registers #0, vendor ID, and #1, version number) command to allow for generic access of vendor-specific information and control.

The initialization group of commands are covered by function calls of an initXX format. These are used prior to establishing a link using the transceiver and can also be used to retrieve link parameters after establishment. The retrieval becomes very important if the device is implementing the G.994.1 handshake procedures.

Statistics are passive in nature. Devices will report events either via interrupts or as bits set in information registers. An LLD can either poll this or service an interrupt associated with the event, and then store or relay the data. The CopperGold™ API provides the statXX set of commands for retrieval of this information in a standard manner (note that this implies that there is a layer of software below the API that is storing events). Some of the types of statistical events can be error counts (blocks or Forward Error Correction errors), loss of signal, frame, or power and information which allows verification of QOS parameters.

The eocXX group of functions are used mainly by ATU-C implementations. This is because almost all eoc commands are initiated by the ATU-C. Even the "dying gasp" command from the ATU-R is generated as a side-effect of local conditions and not as a result of application need.

All APIs will have a "miscellaneous" category (this may be done by default by having a single command in multiple groups). This group of dataXX commands allows a sequence of commands to be initiated and have the results returned after the sequence is finished. For example, a dataXX command to retrieve the far-end vendor ID and version number (eoc READ of registers 0 and 1) will need to be repeated several times to indicate command validity and then the data will come back. It is much better to be able to obtain the information on a single call.

The final grouping of functions for the CopperGold™ API is into the cgXX types of commands that are vendor-specific. It is also possible for other APIs to be created by other vendors. However, in each case, an API must be designed to allow for as much hardware independence as possible to allow for changes in specifications and protocols. For example, the CopperGold™ API that is presently available for ADSL can be directly used for G.992.2 without significant modification. Most changes to the protocol are reduced capacity and the eoc is basically the same. Changes such as making use of the reduced overhead fast/sync combined header byte doesn't change the API command—only how the command is executed or retrieved.

In Chapter 6, we will concentrate on signaling methods, giving both historical information as well as signaling (and routing) options for ADSL configurations.

6 Signaling, Routing, and Connectivity

Signaling (or signalling, to use the British spelling often used in the ITU-T Recommendations) is the process of setting up a connection. This may be a circuit-switched connection, a packet-switched connection, or a cell-relay connection. It is also possible that the "connection" will have already been set up. This type of connection, for various protocols and services, is usually called either a "permanent virtual circuit," in cases where multiple logical connections can take place, or a "semipermanent channel" where the signalling has already taken place (either logically, by use of subscription parameters with the network, or as an automatic effect of powering up a device).

Signaling is normally "state machine based" (discussed in the previous chapter). This means that there are different steps in establishing, or releasing, the connection. There are normally two primary states: idle and active. Each logical connection will start in the idle state and proceed to the active state and then, when released, return to the idle state—often releasing both physical and logical (memory) resources at this time. In the meantime, the transition from the idle state to the active state may have a number of other transitory states. These may include such states as "received all necessary information," "other end notified," "interworking started," "other end requests more information," etc. A connection may succeed or fail. If it fails, then it will normally include a reason in the failure message.

Releasing a connection can be slow or fast. The fast method is basically the same as going on-hook for an analog phone. The user has released the physical, and logical connection and it is up to the network, and the other end, to recognize this fact and to release all other needed resources. Other procedures, particularly digital signaling methods, prefer a more symmetrical release of the call. This basically involves a "handshake" between the two endpoints. The side wanting to release the call initiates the handshake by indicating "ready to disconnect." The network then responds with a "Go ahead and release." The final message is of the nature of "release complete, resources all deallocated." Since releasing a connection is pretty much a definite decision (it isn't usually allowed to have the other end refuse to let the connection be released) error situations normally cause the connection to proceed back to the idle state after some predetermined timeout.

Routing is a matter of recognizing the data frame or packet addresses. The protocol most associated with routing and ADSL is that of the Internet Protocol (IP). In this case, the message will contain a destination address and an origination address. The destination address or "To:" field, indicates where the frame should be routed. The origination address or "From:" field, shows where error messages should be sent or, in the case of requests, where the data, which is specified in the request, should be sent.

Connectivity indicates a complete connection. It may be signaled, routed, or directly connected. The Central Office may provide a direct connection to some other subscribed service (which may be an ISP which will use IP routing or a direct connection to some other data provider). As part of the DSLAM architecture, the primary requirement is that the connection provide adequate bandwidth to support the service. If not, then the network will play the part of the bottleneck to slow down the data services.

6.1 SIGNALING METHODS

Signaling methods fall into three categories: analog, hybrid analog/digital, and digital. The first and last are somewhat self-explanatory, but the second is a bit strange. Hybrid methods were developed as part of the growth of the long-distance digital network. Techniques and equipment were already available for analog signaling so why not adapt them directly for digital use?

Analog methods are basically a variant of "tip" and "ring," which makes sense only if one is familiar with the history of the public telephone system. Instead, think of these terms as "on" and "off" to map them directly to later digital methods. Hybrid methods map these on/off signals to a digital channel (or part of the data channel) but use them in basically the same manner. Digital methods apply framing and packet protocols. This provides a more robust mechanism (what happens if the network just happens to be looking elsewhere at the time you enter a number?) and allows much greater control over the connection and information provided about the connection.

6.1.1 ANALOG DEVICES

Each switch (physical or logical) in North America (it can vary in other parts of the world) directly supports 10,000 lines. This is why North American phone numbers are divided into three segments (plus the "1" for long-distance—which is, not coincidentally, the international code for the North American network). These three segments are referred to as the area code, prefix, and subscriber extension (the latter has different names depending on the reference point). Since the switch can support 10,000 numbers, the extension is limited to the 0 to 9,999 range.

The prefix is sometimes referred to as the local switch number. In other words, prefix 987 may refer to the lines controlled by one switch and prefix 913 may refer to the lines controlled by another switch. With the advent of electronic switches, multiple prefixes may be handled by the same physical switching computer, but logically each prefix is its own switch. A network will have to examine its "routing tables" to determine the switch associated with the prefix. If it is within the same central office location then all such switches (logical or physical) may have direct access to the logical trunk lines between them. If it is outside of the central office location, then a trunk line will have to be used for switching the call. Note that only 1,000 prefixes are possible.

This brings us to area codes. An area code is similar to a prefix except that, rather than identifying a switch, it identifies a region. Once upon a time, the area code almost always meant a "tandem switch"—meaning it was the next step up on the long-distance trunk network. There would be local loops connecting to a central office, then high-frequency trunks between central offices in an area, then lower-frequency trunks between more distant offices in an area. Finally, if connections were needed between areas, they would be routed to a large switch within the area that was connected to other large switches in other areas. This hierarchy can be seen in Figure 6.1. The best reference for the history of the growth of the North American network is "Engineering and Operations in the Bell System" published in 1977 by Bell Laboratories. Unfortunately, it is now out of print, but may be obtained at libraries or out-of-print bookstores.

These three limitations: 1,000 area codes, 1,000 prefixes within an area code, and 10,000 subscriber extensions, indicate a total possible set of numbers amounting to ten billion numbers. This would normally be ample for North America for the foreseeable future. Unfortunately, the total number available is not equally distributed. This has led to an almost continuous addition of new area codes in rapidly growing urban areas.

The first thing that an analog device does for signaling is to go "off-hook." This is the process of taking the receiver off the hook of the phone. The same thing can be done by modems or other equipment (such as fax machines) by interrupting the circuit on the "ring" wire. The central office doesn't monitor the condition of this at all times. Since there are 10,000 potential lines to monitor, dedicated equipment

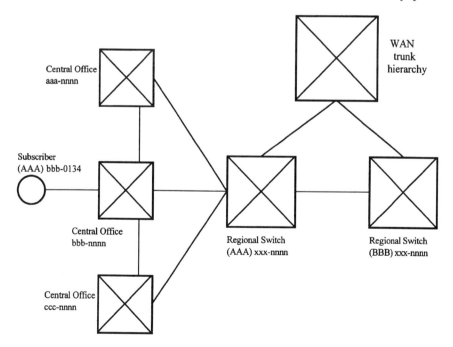

FIGURE 6.1 PSTN switch hierarchy.

to each line would defeat the purpose of traffic engineering and attempts to provide reasonable network access costs.

Instead, what the central office does is to have equipment (physical or a logical process) do a "round-robin" (rather like asking a question of each person in a circle) survey of all (or a subset) of the lines connected to the switch. Once the switchhook has been noticed as being off-hook, equipment is allocated to that line to wait for signaling information and which will provide dial tone. This accounts for the occasional "dial tone" delay when a handset has been picked up. Although pulse dialing is rarely done anymore, the process of signaling is basically the same as with Dual-Tone Multi-Frequency (DTMF) ("push button") phones.

The process is described in Figure 6.2: the subscriber picks up the phone, the off-hook is detected, and dial tone is applied. The user enters the digits or symbols ("#" and "*" on most push-button phones) necessary for the address. The network sends an off-hook to the Terminating CO (TCO). The TCO sends back an off-hook delay-dial message to the Originating CO (OCO) followed by an on-hook start dial message. The OCO then transfers the accumulated address information, the far end gets ringing, and the near end receives "audible ringing." Finally, the far end picks up the phone, the dial tones are removed, and the conversation begins.

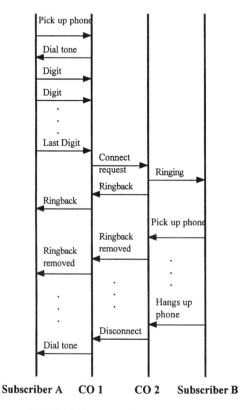

FIGURE 6.2 An analog call scenario.

6.1.2 CHANNEL ASSOCIATED SIGNALING (CAS)

There were two problems with the use of DTMF with long-distance networks. First, it severely limited the number of symbols to be transmitted (to 12). Second, it was easy (once a person had obtained the necessary key combinations) to fool the network and provide network codes via the local phone and confuse the network such that a long-distance call was treated as a local call.

The solution to these problems was to disassociate the signaling from the local phone to the Central Office from the signaling used on the long-distance trunk. Thus, the local phone could *only* communicate with the local switch and the in-band connection was blocked at the central office until the full connection was made. This prevented any illegal misuse of the long-distance signaling network and allowed additional signals to be used.

This could be done in one of two ways. It was either done "in-band" or as part of Common Channel Interoffice Signaling (CCIS). In-band signaling was within the TDM channel on the trunk line. Thus, the signaling could be associated with the affected bearer channel because it was carried on the same channel. In CCIS, the signaling was similar but it was carried over a separate, dedicated, trunk line and the bits were either directly mapped to the trunk line which it controlled or contained mapping information as part of the channel information.

In-band signaling for North American and Japanese T1 digital trunk lines is often called "robbed bit" because the low-order bit of each byte on a digital channel is used for the signaling information. This has the side effect of reducing the data capacity of each bearer channel from 64,000 bps to 56,000 bps (7/8 times 64,000). In Europe, and elsewhere where the E1 transmission system is used, there is no such equivalent data loss because a separate TDM channel is used to carry the signaling information—very similar to the CCIS.

6.1.3 Q.921/Q.931 VARIANTS

X.25 Was the first "modern" signaling protocol which made use of the OSI layer model. In X.25, there were three separate protocol layers: the physical layer, data link layer, and the network layer. The physical layer made use of HDLC frames and the data link layer used a state table which allowed retransmissions and relatively secure transmission of data over what were, at that time, relatively poor transmission lines.

The data link layer, called Link Access Protocol for the B-channel (LAPB) was the basis for a more versatile data link layer protocol, Link Access Protocol for the D-channel (LAPD). LAPD took the basic states for LAPB and added the concept of logical multiplexing, using multiple subaddresses in the address field.

The third OSI layer, the Network Layer, combined two different types of protocols into a single standard. This can be called the Network Layer Call Control and the Network Layer Packet Protocol. As inferred from the name, Network Layer Call Control was concerned with signaling, while the Network Layer Packet Protocol dealt with data transport. Many of the specific architectural, and naming, conventions used in later protocols were missing, but the protocol was able to map directly into the later architectures.

BRI ISDN, as mentioned in earlier chapters, was the protocol which actually used Q.921 and Q.931. This formal architecture allowed multiple use of the signaling channel at the data link layer. It also split the signaling away from the data protocol at layer three. Its main architectural addition, however, was the formalization of primitives across the OSI layer boundaries.

For example, with BRI ISDN, primitives were divided into four main types: Requests, Indications, Confirmations, and Responses. Requests get back Confirmations and Indications require Responses. Interlayer primitives got their own naming conventions with primitives passing between layer 1 and layer 2 called PH_ primitives and primitives passing between layer 2 and layer 3 called DL_ primitives. A formal management entity, for managing system resources, was defined and a system of "planes" for supervisory and control messages was put into place.

At this point of time, the management, supervisory, user, and control messages were still basically a "blob," without any clear layers within the planes. However, Q.921 proved to be sufficiently robust to be used as the basis for Link Access Protocol for MODEMs (LAPM) and Link Access Protocol for Frame Relay (LAPF).

In a similar fashion, Q.931 was sufficient for both BRI and PRI ISDN. It also proved to be a good base for Q.933 signaling protocol for Frame Relay, and Q.2931 signaling for Broadband ISDN (B-ISDN, also referred to as ATM).

6.2 ROUTING METHODS

Routing is composed of two aspects: redirection and encapsulation or protocol conversion. In redirection, a router has access to a "routing table" similar to the long-distance networks. This is basically a table similar to entries in a telephone book. The router receives a frame of a known type (based upon the type of interface line on which the frame has appeared). Since the type is known, it can examine the frame for address information.

The destination address is particularly important. The router can look up the address in the routing table and determine if it is local (the router doesn't need to do anything, as the frame will continue around the LAN) or remote. If it is remote, it will need to search for the location of the destination, how to get to it, and if there is any need of encapsulation or protocol conversion to make the packet compatible with the routed destination.

A special case concerns frames that must be routed to a different local (connected to the router) LAN. Since the router is a *node* (connected device) of each LAN, it has access to all frames that are transported on the LAN. However, LAN A doesn't have direct connectivity to LAN B and, therefore, the router provides a common junction. The router, in this case, only needs to copy the frame from the one LAN to the other (in actuality, acting as the creator of the frame for the second LAN but duplicating information from the original frame). However, (similar to remote locations) encapsulation or protocol conversion may be required if the LANs do not support identical protocols.

We said that the destination address is especially important. However, routers may also perform security checks on frames. This involves the origination address. The router may search its tables for permissions. Can this origination address have access to the destination address? Does it need to perform any authentication procedures (either with the originator or, as part of remote connection, with the destination)?

Assuming that the originator does have permission (or can obtain permission) to access the destination, the router will have to gain access to the remote network. This is likely to involve signaling (by definition, in a sense, since if signaling is not needed then it is not really remote). The signaling method will depend on how the remote network is accessed, but will be involved with one of the signaling methods listed earlier in this chapter.

So, the router has the following duties: look up the destination address; check the origination address and see how permissions match; perform any encapsulation or protocol conversion needed for the frame (the same reverse procedures will be performed on any frames coming back to the originator); copy the frame (after any needed permission procedures or encapsulation/conversion) to a local system or go through signaling procedures to get a connection to the remote network; and start forwarding frames (both directions).

6.2.1 Internet Protocol

Transport Control Protocol (TCP) and the Internet Protocol will be discussed in more detail in Chapter 8. At this time, we will just discuss some aspects of IP addressing. The IP addresses most widely used are from Internet Protocol version 4 (IPv4—usually referred to just as IP). In this version, the frame will have eight bytes (64 bits) of addressing. The first four bytes (32 bits) are the origination address and the second four are the destination address.

The original IP address system allowed for over 4 billion addresses. Even with the huge growth in the use of the Internet and the numbers of PCs within the world, this might still be enough. However, IP version 6 (IPv6) has leapfrogged this limit by increasing address lengths to 128 bits (sixteen bytes) for both the source (origination) and destination addresses.

IPv4 addresses were typically of the type 187.98.203.100 where each byte is a number separated by the "."s. More rarely, the address numbers were represented by hexadecimal numbers such that the above address would be listed as BB.62.CB.64. IPv6 addresses are commonly put into hexadecimal form with 16 bits (two bytes) separated by colons (":"). An IPv6 address might therefore be 1080:CB22:0:0:0:ED:0:1CC4. IPv6 has the added feature where a series of 0s (on 16-bit boundaries) cannot be noted—and they will be extrapolated based on where they must be in the 128-bit series. The above address could be written as 1080:CB22::ED:0:1CC4. Note that the second "0" field cannot be "abridged" as there would no longer be any way to determine just how wide of a field was abbreviated.

IPv6 addresses also have extensions in types of addresses from what was available in IPv4. Some of these are particular address reservations (i.e., Novell's IPX), but others are for particular interfaces. These interface types include unicast, which is unchanged from IPv4 and specifies the specific destination address. Multicast identifies a set of interfaces. This was also available in IPv4.

A new address type in IPv6 is *anycast* which says that the frame may be sent to any one of a particular set of possible hosts. This is rather like tossing the flowers after a wedding. It is up to the host which receives the frame to determine whether to share, or forward, the information. In the case of a "hunt group," it is quite possible that only one of the anycast group needs the information. However, in this case, a router doing IP anycast addresses must have some type of record, or algorithm, to rotate the chosen receiver interface so that one host is not overburdened with anycast messages. Broadcast messages are no longer supported and are replaced (with a more specific set of destinations) by multicast.

The IP address (and other header byte) format is determined by the version number. The version number is located within the first nibble (four bits) of the IP frame. Thus, a router only needs to know what type of a frame (IP, TCP/IP, or PPP-encapsulated TCP/IP) it is to be able to decode the address field properly and do the routing.

6.2.2 Permanent Virtual Circuits

Sometimes the access link to another LAN, or even a remote WAN, may be available on a pre-allocated basis. This is not the same as being a permanent node on both networks, but the effect is similar.

There are two processes needed to access Permanent Virtual Circuits (by a router or by an application). The first is to make sure that the link is initialized. This may be as simple as sending a "hello" message from one endpoint to another. Some protocols may require a more complicated scenario where the two endpoints "reset" each other to make sure that data transfer protocols are synchronized.

The other process is to make a logical connection. A routing table entry is a type of logical connection. The fact that the address is listed (and marked as available) indicates that this particular channel is present. However, the router also needs to know what address, or addresses, are serviced by the connection. This can be done at administration time—saying something like "when logical connection identifier 307 is present, it may be used for destination addresses that begin with IP prefixes 103 and 192." For some types of services, such as Frame Relay, the network can be queried as to what connections are available; however, the logical association must still be known (or administered).

6.2.2.1 ATM Cells

ATM over ADSL will be discussed in depth in Chapter 7. Addressing is part of the ATM Layer, which corresponds to the OSI data link layer (or *is* the data link layer, although it does not have a Q.921/HDLC format). There are two identifiers: the Virtual Path Identifier (VPI) and the Virtual Channel Identifier (VCI). It is important to remember that they really have only *local* significance. In other words, the VPI/VCI is able to be utilized by the network to identify a particular link, but not the end destination.

The end destination is only determined by pre-subscription, or remembered from the signaling setup. Still, the routing tables necessary are basically the same; however, because of the data rates, it is unlikely that routing will be done in the DSLAM. Rather, it will route all ATM traffic to an ATM cell-relay node which will handle actual routing. The DSLAM will provide only ADSL idle bit deletion.

6.2.2.2 Frame Relay

Frame Relay has primarily been used with Permanent Virtual Circuits. These are determined by sending a STATUS_ENQUIRY message to the network node, which then replies with a STATUS message listing all of the PVCs available to the node.

The address field of Frame Relay is a modification of Q.921. It combines the two address fields of LAPD, the Terminal Endpoint Identifier (TEI) and the Service Access Point Identifier (SAPI) into a single field called the Data Link Connection Identifier (DLCI). Frame Relay will be discussed in greater detail in Chapter 8.

Frame Relay has many available options. The Frame Relay Forum has issued a series of "agreements" which specify the list of procedures, parameters, and options that are to be supported in Frame Relay equipment that wants to be able to interwork with other equipment. These agreements limit the DLCI field to a total of 13 bits, at present, which fit the two byte address field of LAPD.

Frame Relay also can support a fuller protocol set. This protocol set is composed of Q.922 at the Data Link Layer and Q.933 at the Network Layer. ITU-T Recommendation Q.922 has an annex (Annex A) which provides the "core" control aspects for PVCs and which basically does the same as the "Unnumbered Information" frame of Q.921. This means that an acknowledgment is not expected and the data link layer is just responsible for sending and receiving the frames. This very basic protocol is useful to make the most efficient use of a data channel. It relies on the fact that most data lines are currently mostly error-free such that any retransmissions or error corrections can be done at the higher layers or applications.

A subset of Q.933 (Frame Relay Forum agreement 4, or FRF.4) allows use of Switched Virtual Circuits (SVCs), but still keeps the same limitations as that used with PVCs—use of Q.922A for data transmissions.

6.3 SIGNALING WITHIN THE DSLAM

The DSLAM provides multiplexed access to a variety of subscriber line protocols and physical line methods. The network-side unit, whether it is an ATU-C or another device, should take care of any "extraneous" material that was needed for proper framing. This leaves the true data heading into the DSLAM.

The DSLAM can act as a router or a switch but, in many cases, it will only route the traffic to an appropriate device. Thus, frame relay data will be routed to a Frame Relay Switch, ATM data will be routed to an ATM cell relay node, and so forth. The important architectural feature is that the DSLAM is able to connect between the subscriber access line and the network service, or endpoint service, needed by the line.

7 ATM Over ADSL

The use of Asynchronous Transfer Mode (AMT) over Asymmetric Digital Subscriber Line (ADSL) is still a very controversial issue although it will probably be settled by the market (which is better than sitting a lot of developers together in a room). The controversy arises primarily out of the desire to make products that interwork. This was true for the T1.413 and ITU-T Recommendations. When DMT was chosen as the "main" ADSL physical line standard, the companies who had proven the feasibility and marketed CAP-based products found they had to supply both types of equipment.

However, the choice of a single physical line standard (DMT) as the primary method makes it much easier for ATU-C equipment to interact with ATU-R equipment, regardless of the manufacturer. The protocol use of ADSL raises the discussion to another level, but it is still an issue very similar to that which was debated over physical line coding. If one company designs an ATU-R that provides PPP frames containing TCP/IP, then it may not be able to work with an ATU-C that expects ATM cells. Providing equipment that is capable of providing a variety of protocols adds complexity and cost—and neither is a marketing advantage to those who want to sell equipment.

ATM adds a high degree of complexity to the ADSL system. First, ATM has a high-overhead structure. This is needed for the types of rates expected for ATM cell relay transfer (25 Mbps is the slowest speed presently defined and ATM25 is expected to be the interface used when ATM use extends to the host operating system) but is, perhaps, excessive when lower rates are involved. Second, ATM is inherently a switched (or cell relay) protocol although most ATM use is still of the "semi-permanent" variety so that no signaling is needed. If the protocol being used is a router-oriented protocol such as Internet Protocol (or the combined TCP/IP) then there is double-layering for connections—a routed protocol encapsulated within a cell-relay switched protocol.

This is fine, if needed; however, that is a source of controversy. *Is* it needed? The primary reason that ATM is so strongly recommended by the standards groups is that it is the current best-proposed replacement for the long-distance network for high-speed data. Use of ATM brings back the switching advantages of BRI and PRI ISDN, but puts the circuits onto the new infrastructure, and this new ATM infrastructure can be traffic-engineered based on new requirements and tariffed according to the capital needs and use. In other words, if ATM is needed for the future, then make it part of the network access standards now so that the people will have equipment which will not become obsolete too quickly.

What about the services that don't need long-distance networks? There will probably be two main choices for ADSL equipment: one will support ATM (probably under the SNAG recommendations as discussed later in this chapter) and the other will support STM and make use of DSLAM redirection to IP routers or direct endpoints. It is possible that ATM-supportive equipment will also support STM since most of the complexity is for ATM support.

7.1 B-ISDN (ATM) HISTORY, SPECIFICATIONS, AND BEARER SERVICES

Asynchronous Transfer Mode, which is a specific implementation of Broadband ISDN, was a later development of the ITU-T (beginning with Recommendations I.113 and I.121 in 1988). Although the original architecture of ISDN kept the borders wide enough to allow both ADSL and ATM, the increase in demand for high-speed data transfer was a surprise to the international standards committees. This was a large part of their change in publication schedules from a four-year cycle to an individual, as-ready, cycle.

Asynchronous Transfer Mode is named because the data frames may occur anywhere within the bit stream (which is why the ADSL Forum has stated that the ATM structure can start at any place within an ADSL frame). A synchronous transfer mode uses a specific format and alignment—that used with BRI ISDN or even that of STM within ADSL. In ADSL STM, the bits are aligned within the ADSL frame.

ATM data are organized into Protocol Data Units (PDUs) which carry their own identification information within the unit. This adds to the overhead (which is considerable), but allows per cell relaying of data. Small cells ease the load on transmission equipment, reduce latency time, and allow a tighter multiplexing on physical circuits. It is always a tradeoff, however, because small data units implies a higher percentage overhead.

7.1.1 Broadband Bearer Services

In the discussion on BRI ISDN, we talked about bearer channels (B-channels) which provide bearer services. These services differ between Narrowband-ISDN (N-ISDN) and Broadband-ISDN (B-ISDN) *only* inasmuch as certain services require the speeds available with ATM or B-ISDN. It is fully possible to use ATM to provide the same services as are done with N-ISDN but, in many instances, the extra speed is unwarranted. BRI ISDN can support 128 kbps (uncompressed) data flow for the Internet using MultiLink Point-to-Point Protocol (ML-PPP). ADSL, using a TCP/IP over PPP (*not* ML-PPP) in STM can provide up to 8 Mbps downstream access to the Internet (depending on bottlenecks). Full STM-4 ATM can theoretically supply amounts of data that no present-day computer system could handle; however, they all allow use of the same service—only the speed varies.

So, there are three general categories of differences in the bearer service capabilities between N-ISDN and B-ISDN. First are services such as circuit-switched voice for which N-ISDN has sufficient speed and, within which, B-ISDN speeds are excessive. Second are services, such as PPP (or ML-PPP) supporting Internet access protocols that can be provided by both classes of systems but which can, at least in theory, benefit from the higher speeds of B-ISDN. The third category includes services which are impossible to support at N-ISDN speeds—video services such as MPEG-II video feeds or HDTV channels.

The ITU-T Recommendations concerning Broadband ISDN (B-ISDN, category 3 as discussed above) are called the F.2xx series of Recommendations (the "xx" is replaced by particular number). General aspects of vocabulary, functional aspects, and architectural components of B-ISDN are located in ITU-T Recommendations I.113, I.211, I.121, I.311, I.321, I.327, and I.150 (I.432 is discussed in the following section). They are concerned with the physical layer. I.431 Provides reference models as concern the UNI and ITU-T Recommendation I.460 discusses minimum functions needed by customers in order to make use of the various ATM layers.

Service aspects of B-ISDN are discussed in ITU-T Recommendation I.211. The two primary service categories are interactive and distribution services. The three subdivisions of interactive services are: conversational, messaging, and retrieval services. Distribution services are aligned according to the amount of control the user has concerning the information flow.

7.1.2 Specific Interactive and Distribution Services

The three subdivisions of interactive services are distinguished by the amount of available real-time dependency. Conversational services, the first subdivision, allows bidirectional communication in a real-time situation. I.211 mentions high-speed data transmission, video conferencing, and video telephony (and video surveillance) in these categories. Most services in this subdivision are video since there are few other services which require both the speed and the immediacy of this category.

Messaging services, the second subdivision of interactive services, are no longer real-time. This involves an aspect of real-time in that services are expected to be created on a real-time basis, but stored and then retrieved at a later time. Video and mail message services, which may include HyperText Markup Language, HTML, hyperlinks, fall into this category.

Retrieval services comprise the last subdivision in the interactive services category. This is more in the nature of an archival situation. Movie libraries or high-density digitized artwork or sets of encyclopedias might be accessible as a retrieval service with a public or private library.

Distribution services are similar to cable broadcast services. Interactive services entail linking data to specific requests. Distribution services allow the data to always be available. The two specific subtypes are concerned with the control that the user has over the data. A set of radio or video (HDTV, perhaps) stations that are always putting out specific programs is an example of the first. If the ability to start, stop, or change the location (beginning, end, or somewhere in between) is present then the service falls into a second class. Note that this differs only slightly between interactive retrieval services; with interactive retrieval, the user has a choice about what data are presented and, with class two distribution services, there is no choice about *what* data but only what portion of data is to be presented at the time.

7.2 B-ISDN OSI LAYERS

In Chapter 4 we briefly discussed the Open Systems Interconnection model. Figure 7.1 shows the B-ISDN protocol model associated with the OSI model. The ATM physical layer (which would be ADSL or ATM25) occupies the first layer. The ATM layer is parallel to that of the Data Link Layer for N-ISDN. FInally, the ATM Adaptation Layer (AAL) is equivalent to the Network Layer

Since we will not discuss N-ISDN in-depth, it may be useful to summarize some of the general ISDN architectural aspects. In addition to the OSI layers, the ITU-T indicates that three types of "planes" are useful for most protocols. These planes are the User plane (U-plane), Control plane (C-plane), Supervisory plane (S-plane) and plane management functions.

These planes are primarily defined in terms of the *Integrated* part of Integrated Services Digital Network. In an integrated environment, it is possible (even likely) that the same protocol environment will be used for more than one service—and perhaps more than one at the same time. Whenever such parallel activity occurs, it is necessary to keep the instruction and data path through the layers directed to the right functions and entities. This is the responsibility of the C-plane. The S-plane is delegated to act as the "watchdog" for the system and provides error logging, statistical grouping, and possible reinitialization when needed.

The U-plane is just a fancy way of saying the individual application connection's route through the protocol stack. Plane management functions have changed over the years as the ITU-T gets feedback from the real world as to what is needed. Management functions deal with system resources such as timers, buffers, possibly hardware (when swappable), etc. Early ITU-T Recommendations treated the management plane functions as a global entity. Today, however, as Frame Relay, ATM, ADSL, and other more versatile standards are increasingly available in consumer markets, the management plane really falls into sub-layers according to the OSI layer model. Thus a management primitive may go from the physical layer to the management entity. This, in turn, causes an interlayer management entity to be generated to pass needed information from one layer to another.

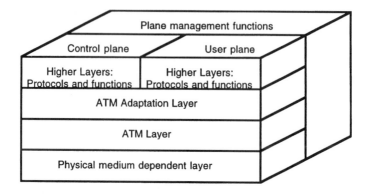

FIGURE 7.1 B-ISDN (ATM) protocol model. (From ITU-T Recommendation I.121.)

7.3 ATM PHYSICAL LAYER

The ATM physical layer is defined by ITU-T Recommendation I.432. This is for the User-Network Interface (UNI) as opposed to the Network-Network Interface (NNI). Although, as mentioned previously, the physical aspects of data transmission at ADSL rates must be done (after control registers are properly initialized) by hardware support, the general aspects of the physical cell structure will be of use in understanding the methods of structuring the code.

Transport of ATM over ADSL will use a similar physical structure as is defined in I.432, but the ATM cells (discussed in the next section) will likely be put into this format upon entry into an ATM cell-relay network. The ADSL Technical Report 2 (TR-002) called "ATM over ADSL Recommendations" states that the cell octets may be transmitted aligned to any bit within the ADSL octets. On the receiving side, there should be no assumption of alignment. This basically means that all bits must be examined until the framing pattern is seen, as described by the 'A1' and 'A2' bytes in the ATM block header shown in Figure 7.2.

This figure shows the STM-1 bit interface which is used over a bidirectional 155.520 Mbps interface. This is known as Synchronous Transfer Mode 1 (STM-1, note that, although the acronym stands for the same thing, the usage is different than STM for ADSL). I.432 also partially defines an STM-4 mode that supports 622.080 Mbps. STM-1 is one of the formats capable of being used within the ATM cell-relay network. Synchronous Optical Network, (SONET) is another highly related non-electrical format.

The ATM cell, which is of greatest importance for ADSL transport, is listed only as a 53-octet (8-bit byte) unit of data. This means that, of the 2,430 bytes within an STM-1 "container," there are 90 bytes Section OverHead plus 221 (average, calculating 5/53 of a 2340 STM-1 payload) other overhead bytes. This is a 12% overhead for the data within the ATM cells—not including the ADSL frame overhead. A Frame Relay PVC, using Q.922A, with a 1,500-byte frame only has about 0.33% overhead.

Each of the layers, within the B-ISDN architecture, is further split into functional sections as may be seen in Table 7.1. The Physical Medium (PM) sublayer of the physical layer can use any physical interface—including ADSL. Bit timing is part of the Network Timing Reference (NTR) requirement for ATM use of ADSL.

The Transmission Convergence (TC) sublayer must be adapted to the needs of the PM sublayer. Thus, the ADSL Forum, ANSI, and ITU-T documents must indicate just how the ATM TC is to be done within the context of the ADSL physical line environment. Many of the generic requirements of the TC sublayer are addressed in ITU-T Recommendation I.432. Both ANSI T1.413 and ITU-T G.992.1 and G.992.2 directly address application of the TC sublayer to the needs of ADSL. The Transmission Frame Adaption function, in particular, is associated with the placement of the STM-1 payload into the ADSL frame.

7.4 ATM LAYER

The STM-1 (or other) structure acts as a "container" for a set of ATM cells; however, it is the ATM cells which have the data that can be used for different purposes. Each

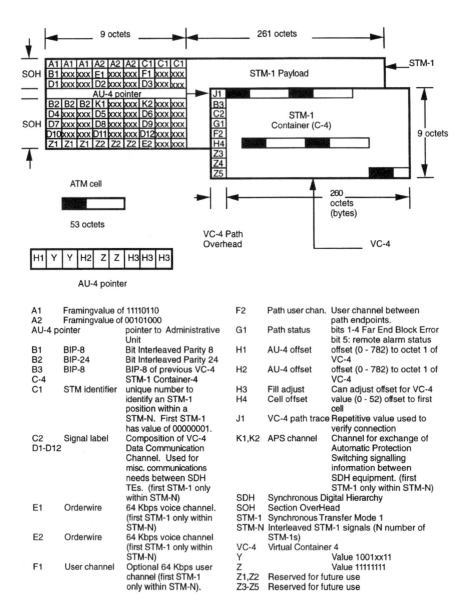

FIGURE 7.2 ATM cells mapped into the STM-1 signal. (Adapted from ITU-T Recommendation I.432.)

ATM cell is composed of 53 bytes. The first five bytes are a header (providing information type), address information, priority, add Header Error Control (HEC) (providing an approximate equivalent to the FCS of a BRI ISDN frame). Actually, because of the relative size (7 bits for 48 bytes of data), the HEC can provide a small degree of error correction as well as detection. Whether this is useful or not

TABLE 7.1
B-ISDN Layer Functions

	Higher-Layer Functions		Higher Layers	
	Convergence	CS	AAL	
	Segmentation and reassembly	SAR		
	Generic flow control		ATM	
	Cell header generation/extraction			
	Cell VPI/VCI translation			
	Cell multiplex and demultiplex			
Layer Management	Cell rate decoupling	TC	Physical Layer	
	HEC header sequence generation/verification			
	Cell delineation			
	Transmission frame adaptation			
	Transmission frame generation/recovery			
	Bit timing	PM		
	Physical medium			

Source: From ITU-T Recommendation I.121.

is dependent on the application; the act of making use of the HEC to correct errors (on *every* cell) may add more latency than retransmitting every once in a while.

7.4.1 ATM CELL FORMATS

Figure 7.3 shows the User-to-Network Interface (UNI) for an ATM. Figure 7.4 indicates a Network-to-Network Interface (NNI). Note that the only real difference between the two are that the Generic Flow Control (GFC) field from the UNI header is replaced with an additional four bits of Virtual Path Identifier data. This is because of the specific needs of the two interfaces. User equipment (say an ADSL user transporting ATM data, to keep it directly relevant) needs to maintain a degree of flow control between the network and itself. The network, however, will be dealing with nodes that have (if well designed) equal data capacity and are dedicated to data routing and transport. It is more important that the network have access to additional "paths" (equivalent to TDM channels in BRI ISDN) between nodes.

An ATM cell is transporting data, however, the purpose of the data is not fixed. It may be used to convey information about the physical layer, the ATM layer, or the ATM Adaptation Layer (AAL). The Cell Loss Priority (CLP) bit of the fourth byte of the cell header is defined as a indicating eligibility—to be discarded if necessary. However, Recommendations I.361 and I.150 do not fully agree. In practice, the bit is used to indicate use of the cell for physical layer needs as seen in Table 7.2.. If the cell is used for physical layer Operation And Maintenance (OAM) then it is *not* passed on up to the ATM layer.

8	7	6	5	4	3	2	1	Bit / Octet
GFC				VPI				1
VPI				VCI				2
VCI								3
VCI				PT		RES	CLP	4
HEC								5

CLP Cell Loss Priority
GFC Generic Flow Control
PT   Payload Type
RES Reserved
HEC Header Control
VPI Virtual Path Identifier

FIGURE 7.3 B-ISDN UNI format. (From ITU-T Recommendation I.361.)

8	7	6	5	4	3	2	1	Bit / Octet
VPI								1
VPI				VCI				2
VCI								3
VCI				PT		RES	CLP	4
HEC								5

CLP Cell Loss Priority
GFC Generic Flow Control
PT   Payload Type
RES Reserved
HEC Header Error Control
VPI Virtual Path Identifier

FIGURE 7.4 B-ISDN NNI format. (From ITU-T Recommendation I.361.)

The OAM description listed in I.610 are named as Flow (F) points. F1 through F3 are used for network line monitoring, while F4 and F5 are used for end-to-end checking of the integrity of the path. In other words, the data will flow from one

TABLE 7.2
B-ISDN UNI Preassigned Cell Header Values

	Octet 1	Octet 2	Octet 3	Octet 4	Octet 5
Reserved for use of the physical layer	PPPP0000	00000000	00000000	0000PPP1	HEC
Unassigned cell identification	AAAA0000	00000000	00000000	0000AAA0	HEC

Source: From ITU-T Recommendation I.361.

end to another. Use of the Fx OAM messages can determine reliability of specific parts of the data flow path. F1 is the regenerator section, F2 is the digital section and F3 is the transmission path. F4 and F5 refer back to the items mentioned in Figure 7.3 as the Virtual Path Identifier (VPI) and Virtual Channel Identifier (VCI). The VPI (equivalent to a TDM channel for N-ISDN) is the large "pipe" within which the VCIs are routed.

7.4.2 VIRTUAL PATHS AND VIRTUAL CHANNELS

There is a hierarchy on the data paths for ATM. The actual data paths from one endpoint to another are considered to be Virtual Path Connections (VPCs, identified by Virtual Path Identifiers, VPIs). Within this "virtual" path (virtual because in a cell relay system the path is rarely the same for the duration of the connection) there may exist one or more Virtual Channel Connections (VCCs) which are identified by Virtual Channel Identifiers (VCIs). ITU-T Recommendation I.150 discusses some of the more specific aspects of VPCs and VPIs.

Since a VPC will not necessarily occupy the same data path for the duration of the connection, it is unwieldy for each node in the network to know the VPI. Furthermore, every node would have use the *same* set of VPIs so that only completely disjointed networks could use the same VPIs for different paths. A limitation on the network to a total of 4096 (2^{12} possible NNI VPIs) data paths is not only unacceptable, it is unnecessary.

What is done is that the VPI has only *local* significance. When a data path is set up (by whatever mechanism), a VPI is allocated which will identify the VCC between the two nodes (perhaps the user and network—which is limited to 8 bits or 256 values). However, a new (and possibly different) VPI will be allocated between each set of nodes as the path is established. If a "lap" (connection between two nodes) is broken, all that is required is for the nodes to renegotiate a path between themselves—no other nodes in the VPC need know the new VPI.

This infers that each cell-relay network requires a central administration node which may recognize the specific nodes on a particular connection, in addition to the duration, and other management and administrative information. It is not practical for each node to have full address and accounting information because the nodes have knowledge only about the links that they have to other adjacent nodes.

There are two main restrictions on how the full VPC is set up. The first is that it must be able to support the Quality Of Service (QOS) that has been requested (and paid for) and the second is that cells be relayed in a sequential fashion. This second requirement implicitly indicates that there will only be *one* physical data path for each VPC (although the exact route may change over the duration of the connection). If more than one physical data path was in use between two nodes in a VPC, there would be no way (without modification of ATM cell contents) to make sure that the cells, arriving wherever the data paths reconnected to the same node, would stay in the same order as transmitted.

The VCC can be set up as semipermanent (ie., subscription such that the virtual channel is present whenever the physical layer is active and the encompassing VPC is present). A signaling VCC may also be set up using "metasignaling" procedures. Use of this signaling VCC can set up additional data-oriented channels (as will be seen in the discussion on Q.2931).

The cell multiplexing and demultiplexing function shown in Table 7.1 is the process of ensuring the appropriate VPI/VCI is used for transmitting a particular data stream and mapped to the right data stream on reception. Occasionally, the process of translation is mapped into an appropriate Service Access Point (SAP). This is particularly true for Q.2931 signaling on ATM, where the VPI/VCI maps into a signaling route (which probably will not be routed directly to higher protocol layers).

Cell-header generation and extraction is concerned with the five header bytes of the ATM cell. Although the HEC byte is really generated by the TC sublayer of the physical layer, there must be a byte left in the header to allow insertion. The higher layer (AAL) puts in the other four bytes (on transmission) to enable the cell to be correctly routed. Generic Flow Control makes use of the GFC nibble on the UNI ATM cell header in a manner not fully explained as of yet.

7.5 ATM ADAPTATION LAYER

Figure 7.5 shows the service classifications for the ATM Adaptation Layer (AAL) according to ITU-T Recommendation I.362. The AAL is not exactly a layer three protocol. It provides enhanced features for the ATM layer and access by the higher layers. The two sublayers, Segmentation And Reassembly (SAR) and Convergence

	Class A	Class B	Class C	Class D
Timing relation between source and destination	Required	Required	Not required	Not required
Bit rate	Constant	Variable	Variable	Variable
Connection mode	Connection-	Connection-	Connection-	Connectionless

FIGURE 7.5 AAL Service classifications. (From ITU-T Recommendation I.362.)

ATM Over ADSL

Sublayer (CS) are present. The SAR is responsible for converting between the data form needed by the higher layers and the 48-byte ATM cell contents. The CS is application specific and may vary depending on the application and specific AAL type chosen.

These four classes of service correspond to particular AAL protocol types. Constant Bit Rate (CBR) services are supported by AAL type 1. This is defined by Class A. AAL types 3 and 4 provide the support needed for connectionless (CL) services (Class D). Class C services may use AAL types 2 and 3. AAL type 5 was added to the list by ANSI because they saw a need for the ability to transparently support higher-level protocols without AAL intervention.

The AAL types are shown in Figure 7.6. Since the ADSL specifications indicate that types 1 and 5 will be primarily transported over ADSL, we will concentrate on them. In-depth explanation of the other three types is left to readers who want to better understand the underlying ISDN definitions and services. Note that the original AAL type 4 was merged into AAL type 3 because the fields overlapped; type 4 is only slightly more specific in its use of the "reserved" 10 bits that were unused in type 3.

AAL type 1 mentions three fields: Sequence Number (SN), Sequence Number Protection (SNP), and SAR-PDU payload. The SN is a modula 16 number allowing a degree of sequence checking to ensure that no cells have been lost or routed out

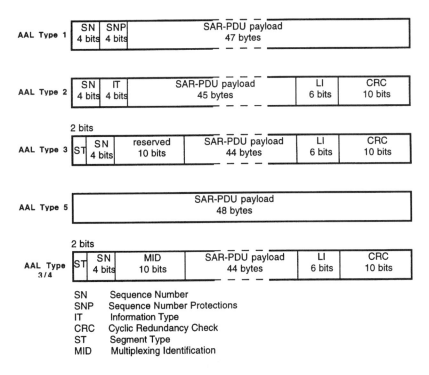

FIGURE 7.6 AAL Types. (From ITU-T Recommendation I.363 and ANSI T1S1.1.)

of order. The SNP gives the possibility of additional error checking within the PDU (or just the SN). This is at the AAL, rather than the HEC at the ATM layer.

7.5.1 AAL Type 1

Class A indicates that the service will be bit-timed, at a constant rate, and with a specific connection. The SN and SNP fields preceding the data indicate that there should be some mechanism to report to higher layers of erroneous cells that were received. Also, the class implies that a loss of synchronization requires handling.

7.5.2 AAL Type 5

Figure 7.7 shows the *last* CS-PDU for a type 5 set of cells. Since AAL type 5, handling Class C services, is basically a transparent passing of data from the SAR to the CS, it means that most of the general responsibilities of the SAR sublayer are left to the CS. Thus, the suffix of six bytes at the end of the last cell provides knowledge of just how much of the X times 48 length data frame is useful data and an HDLC-like Cyclic Redundancy Check (CRC) is provided for error detection. Sometimes this AAL type is called the Simple and Efficient Adaptation Layer (SEAL)—it tries to lower the overhead necessary when working with ATM.

AAL type 5 gives up types 1 through 4 capability of checking individual cell errors but allows for a greater check for the entire sequence. Note that all other cells prior to the last cell must either have a full 48 bytes of data (preferred) or the end cell must be marked as last (possibly by setting the PT field of the ATM cell).

The total amount of data allowed in a sequence of AAL type 5 cells is 64K (65,536) determined by the two-byte length field available in the CS-PDU suffix. Various optional sublayers have been proposed for use in conjunction with AAL type 5-specific sublayers based upon the type of data being conveyed. One of these is the Frame Relaying Specific Convergence Sublayer (FRCS), designed to help in interworking between ATM and Frame Relay (see Chapter 8). Another is called the Service-Specific Connection-Oriented Protocol (SSCOP) for B-ISDN signaling. Table 7.3 gives a summary of the various AAL types (including types 2 and 3/4).

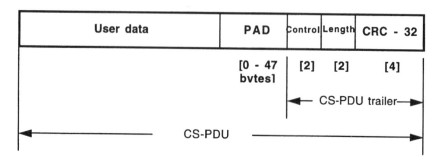

FIGURE 7.7 Last payload for AAL Type 5. (From ANSI T1S1.1.)

TABLE 7.3
ATM Adaptation Layer (AAL) Protocol Types

	Service Provided	Functions in Type	SAR Sublayer	CS Functions
Type 1	Transfer of SDUs with constant bit rate. Timing info indication of lost or errored unrecoverable info.	SAR cell delay handling lost and misinserted cells source clock recovery monitoring and handling of AAL-PCI errors monitoring and possible corrective action for user information bit errors.	For further study.	High-quality audio and video may have forward error correction. Clock recovery capability Time stamp services Lost and misinserted cells handled
Type 2	Transfer of SDUs with variable bit rate. Timing information indication of lost or errored unrecoverable information	SAR cell delay handling lost and misinserted cells source clock recovery monitoring and handling of AAL-PCI errors monitoring and possible corrective action for user information bit errors.	For further study.	High-quality audio and video may have forward error correction. Clock recovery capability. Lost and misinserted cells handled.
Type 5	Efficient and versatile CS handling.	CRC checking. Ability to handle variable data rates. Control field for future use.	None	Error checking. Retransmission. Flow control. SAR. Pipelining. Data link failure detection. Signalling (SSCOP).
Type 3/4	Message-mode service. Streaming-mode service.		Preservation of CS-PDU. Error detection. Multiplexing and demultiplexing.	Preservation of AAL-SDU. Mapping between AAL-SAPs and ATM. Error detection and handling. SAR. Info identification. Buffer allocation size.

Source: From ITU-T Recommendation I.363 and ANSI T1S1.1.

7.6 ATM SIGNALING

The primary method of signaling within ATM is governed by ITU-T Recommendation Q.2931. At the Network Layer, Q.2931 is very similar to N-ISDN in the general form, state structure, and use of message types and Information Elements (IEs). Of course, it also makes use of the ATM Layer at the "link layer" or OSI Layer 2.

7.6.1 Lower Layer Access

Q.2931 makes use of four primitives in its interactions with the ATM Layer. These four primitives are: "AAL_ESTABLISH," "AAL_RELEASE," "AAL_DATA," and "AAL_UNIT_DATA." These are parallel in purpose and effect to the N-ISDN primitives between the signaling entity and the data link layer. As is true for most ITU-T primitives, there are four types of primitive interactions: requests, indications, responses, and confirmations. For Q.2931, responses are not actively used. Thus, any indication is an autonomous operation (no response required). However, AAL_ESTABLISH and AAL_RELEASE requests need to have confirmations to ensure that the link condition is as desired.

Q.2931 makes use of similar, but not identical, terminology concerning the Virtual Path and Virtual Connection. An additional level of indirection is brought into the documentation. This is the Virtual Path Connection Identifier (VPCI); in many cases, this is identical to the VPI. However, in the case of "non-associated" signaling, it is possible to use the signaling channel on other VPIs. Thus, the use of a VPCI allows a level of indirection to indicate another VPI whose specific value may not be known to the signaling channel on the signaling VPI.

The "main" VPI must be in place before any signaling can take place. As of the February 1995 issue of Q.2931, the metasignaling mechanism to set up a signaling channel was still not defined. However, use of the AAL_ESTABLISH_REQUEST primitive still makes the signaling connection viable. The VCI for the signaling channel is presently fixed at the value of 5, therefore a combination of VPI/VCI of xxx/5 will be mapped to a signaling Service Access Point (SAP) and be directed to the Q.2931 signaling module. This is very similar to the SAPI 0 for use within Q.931 on N-ISDN applications.

7.6.2 General Signaling Architecture

B-ISDN signaling was generated because it was thought likely that B-ISDN would need to interact with N-ISDN on a regular basis. Therefore, there is a mapping of certain elements (described later in more detail) between the B-ISDN and N-ISDN signaling protocols.

Each Network Layer entity encompasses a specific state table. As discussed in Chapter 5, a state machine has three main components: state, events, and actions. The state is composed of two parts: the general state (what has generally happened before and what is allowable to happen) and specific state information that is retained from interactions occurring from events in previous states. Events may be ITU-T primitives (ATM_CONNECT_REQUEST from higher layers, AAL_RELEASE_INDICATION from the lower layer, MAAL_TIMER_RESPONSE

ATM Over ADSL   121

from the management entity, etc.) or they may be other application-defined primitives. At any rate, an event is something that is triggered by an outside layer or hardware event. Actions are the result of an event in a specific state. Often, part of the set of actions will be a "transition" to a different state.

7.6.2.1 User-Side States

There are 11 states required for the User (terminal) side of the UNI for Q.2931. There are also an additional two states that may be desirable for interworking (particularly with N-ISDN), but are not mandatory if interworking is not needed or if the interworking network is known not to require the service. These two states are called "overlap sending" (U2) and "overlap receiving" (U25). They are needed when an application, or network node, requires connection, but doesn't know all of the addressing information needed to complete the connection. Thus, U2 is the equivalent of the terminal going off-hook has been detected but ringback has not been applied, to use an analog signaling description. For most PRI ISDN and, presumably, non-interworking B-ISDN, all connection information will be known before the connection is attempted. U25 is from the opposite direction—where the network wants to make a connection within the endpoint controlled by the terminal equipment (multiple links are always possible in ISDN and these may either be logical or physical links) but does not know the full address at the time. This is rarely seen in any of the ISDN environments but is possible for interworking situations where "pass-through" signaling is available. An example of this might be QSIG support by Private Branch Exchanges (PBXs) or Private Integrated services Network eXchanges (PINXs).

The states that are more universally needed include "null" (or "idle") (U0) where no call exists and all data members have their default initial values. After an outgoing call request is sent, the state proceeds to the "call initiated" (U1) state. Depending on whether all destination address information has been sent, the state table will then proceed to either U2 or "outgoing call proceeding" (U3). Once the other end has been offered the call (to accept or deny) and this has been indicated to the originator, the state proceeds to "call delivered" (U4). From this point, the other end can accept the call and finish in "active" state (U10) or it can deny the call and go to state "release indication" (U12). It is also possible for the other side to refuse the call at any other state. If it occurs in U1 state, then there is no need to proceed to U12 as there is no indication that there is any need to reinitialize the destination link—only tell the network that the link is no longer needed.

A similar pattern happens on incoming calls. An incoming call request causes a transition to "call present" (U6). The user-side has a couple of options at this point. Remember that there are three possible events that trigger outgoing state changes. These are notification that all information needed has been obtained, that the other end is being notified, and that the other end has accepted (or denied) the call. For the user-side, any one of these situations may happen and does not *require* a progression from one state to the next.

There are three equivalent states to these events: notification that all needed information has been obtained corresponds to user state "incoming call proceeding"

(U9), alerting the local user is shown by the state of "call received" (U7), and accepting the call is defined by entering "connect request" (U8) state. So, a transition from U6 to U8 is possible if the user application layer deems it desirable. However, it is also possible to go from U6 to U7 to U8 (i.e., on a speech service call if the "telephone" rings) or from U6 to U9 to U8 if some other higher layer needs to determine suitability but no explicit alert is needed. The transition to state U8 is slightly different from the change to active for the outgoing call states because the call is not truly "awarded" until the network has acknowledged that the user-side has accepted the call. This is when the transition to U10 takes place.

Disconnection (if not done in the initial states) involves an exchange of messages between the user-side and network-side. In other words, there are two separate situations that is a bit different from the analog world. These situations are "disconnect" and "release." Disconnect means that the call is no longer desired; release means that the call no longer exists. In the analog world, the two are synonymous because "hanging up" the phone is a unilateral action. With an integrated network, it is considered "proper" to coordinate the release of resources which is involved with the release of a connection.

This requires a transition to "release request" (U11) when the user tells the network it wants to release the call (and is thus in a disconnected state) and awaits the network to confirm this before returning to U0. For a release initiated by the network, the state progresses to "release indication" (U12) and the transition to U0 occurs when the higher layers agree to the release (or when a timer event happens indicating that the network should not wait any longer).

7.6.2.2 Network-Side States

The network-side states are parallel to those of the user-side. In other words, when the user-side makes a transition to state 1, the network-side is expected to make the same transition. This transposes "outgoing" and "incoming;" however, it helps tremendously in ensuring that the states of both sides are coordinated. Note that because of "automatic" actions (times when events are assumed rather than explicit), a transition may be made so that the in-between "matched" state is only a logical transition (the state is not in the state long enough to allow further events in that state).

Network-side has state "null" (N0) which is identical to that of the user-side. When the user-side requests a connection and transits to state 1 (U1), the network-side, upon receiving the connection makes the change to "call initiated" (N1). If more information is needed, a message is sent back to the user to notify it of this condition (which will call it to change to state U2) and go to the "overlap sending" state (N2). Otherwise, if all needed information has been received, it goes to "outgoing call proceeding" (N3). When the network has decided that the call is proceeding (because of a primitive received from another network or administration tables), it tells the user and goes to "call delivered" (N4). Finally, when the network is waiting for the response on the incoming call request (outgoing for the user—and outgoing for the network if it must pass the request along), it goes to state "call present" (N6). Once the call has been accepted, it will go to "active" (N10).

A similar situation occurs for "incoming calls" (outgoing for the user-side). In each instance, the state transition parallels what is happening to the user-side. One of the reasons for this parallel state mapping is that the network-side is likely to be connected to other network nodes, in which case it must perform as a network-side *and* a user-side. Keeping the states parallel allows the network to synchronize the behavior for the different line interfaces (say that the incoming connect request is on Line 1 and the interworking message goes out on Line 2). In the next few sections, we will examine the types of messages sent within the AAL_DATA_INDICATION and AAL_DATA_REQUEST primitives. Note, however, that other events will always be possible.

7.6.3 B-ISDN Message Set

Q.2931 gives slightly different message sets that are needed for specific configurations. For B-ISDN call and connection control, the messages PROGRESS, SETUP_ACKNOWLEDGE, and INFORMATION are not required. These messages are only used for interworking situations. Table 7.4 lists the combined message set for these two situations. SETUP_ACKNOWLEDGE is needed to indicate possible transition to state 2 ("overlap sending") and PROGRESS is needed to inform the endpoints (including intermediary network nodes) of the progress of the connection request, or interworking requirements that may affect the final disposition of the connection. INFORMATION allows N-ISDN information to be carried through the B-ISDN.

A Q.2931 message has three required fields, just as Q.931 does; however, the contents of these message fields vary a little from that used within N-ISDN largely because of the different needs of the different networks. The four required fields are: the Protocol Discriminator, the Call Reference, Message Type, and Message Length. For N-ISDN, the message length is not always needed.

The Protocol Discriminator indicates just how the message is to be interpreted. A value of 8 is used for N-ISDN Q.931. A value of 9 is used for B-ISDN Q.2931. The fact that this first byte indicates the syntax for deciphering the message is very important. Only a fixed location within a message can be used for multiple interpretations, otherwise, there is no way to determine just *how* to decide on the proper protocol. Designating the first byte (similar to that done within the Internet Protocol version number) makes it easier and allows for more flexibility on the information format.

The Call Reference is composed of three fields: the length field (contained in the lower-order nibble of the first byte), the call reference flag (which indicates which side allocated the call reference value, indicated by having a '0' in the upper bit of the second byte), and the call reference value which, for B-ISDN, is normally three bytes. This allows for the possibility of 8,388,608 (2^{23}) possible Call Reference Values (CRVs). N-ISDN, in contrast, allows up to two bytes for the CRV, but rarely uses more than one for BRI (many PRI ISDN lines use two). Actually, only 8,388,606 values are available because the values of 0 and 8,388,607 (all '1's) are preallocated by the protocol. Zero indicates a "global" CRV and all ones indicates a "dummy" CRV. These special CRVs are used within certain messages and contexts. Presently, the global CRV is used with RESTART message procedures. The dummy CRV is expected to be used with supplementary services at some point (supplementary services are additional features not directly related to basic connection setup).

TABLE 7.4
B-ISDN Message Types

Message Type (byte 1)
Bits

8	7	6	5	4	3	2	1	
0	0	0	0	0	0	0	0	Escape to nationally specific message type (see Note 1)
0	0	0	-	-	-	-	-	Call establishment messages:
			0	0	0	0	1	— ALERTING
			0	0	0	1	0	— CALL PROCEEDING
			0	0	1	1	1	— CONNECT
			0	1	1	1	1	— CONNECT ACKNOWLEDGE
			0	0	0	1	1	— PROGRESS
			0	0	1	0	1	— SETUP
			0	1	1	0	1	— SETUP ACKNOWLEDGE
0	1	0	-	-	-	-	-	Call clearing messages:
			0	1	1	0	1	— RELEASE
			1	1	0	1	0	— RELEASE COMPLETE
			0	0	1	1	0	— RESTART
			0	1	1	1	0	— RESTART ACKNOWLEDGE
0	1	1	-	-	-	-	-	Miscellaneous messages:
			1	1	0	1	1	— INFORMATION
			0	1	1	1	0	— NOTIFY
			1	1	1	0	1	— STATUS
			1	0	1	0	1	— STATUS ENQUIRY
1	1	1	1	1	1	1	1	Reserved for extension mechanism when all other message type values are exhausted. (See Note 2)

Note 1: When used, the message type (excluding the message compatibility instruction indicator) is defined in octet 10 of the message and the contents follows in the subsequent octets, both according to the national specification.

Note 2: In this case, the message type (excluding the message compatibility instruction indicator) is defined in octet 10 of the message, and the contents follow in the subsequent octets.

Source: From ITU-T Recommendation Q.2931.

The message type is that listed within Table 7.4. The message type indicates just what type of event the message is meant to convey to the protocol. The message type field, as seen in Figure 7.8, for B-ISDN has two bytes. The second byte is to allow for the overriding of "normal" behavior upon errors. If bit 5 of byte two of the message type field is set to '1' then the lower-order two bits are used to say whether to clear the call, discard and ignore, or discard and report if an error is encountered in the coding of the message. The next two bytes are the message length.

7.6.4 Information Elements

A message content field is made up of a series of zero or more Information Elements (IEs). The IE has four fields. These are the IE identifier, the IE explanation field, IE length field and the IE contents field (if needed). Figure 7.9 shows a "generic" IE format. The "flag" and "IE action ind." fields are used similarly to the same named fields within the message type, except that the special actions can apply either to the IE or the entire message. The coding standard is basically either the ITU-T, ISO/IEC, a national standard, or a standard defined for the network-side (for interworking). Generally, the coding standard (a value of '00') should be used to indicate ITU-T coding unless there just isn't anything within the ITU-T standards that can be used.

Information elements have three possibilities for extension. One is use of an alternate coding standard. This, however, must be known as a distinct standard. This implies that equipment designed for North America (for example) will not work within the European Community if national standards are used. A second method is to use the "IE extension value" which uses a value of 255 (8 '1's) in the IE identifier field and which, when used, allows a two byte IE identifier to be specified.

The third possibility enters the realm of the "codeset." A codeset indicates that a certain mapping between a symbol and a meaning exists. (This is also relevant to cryptography and cyphers.) The default codeset within Q.2931 (and Q.931) is codeset 0; however, it is possible to *change* codesets, either temporarily or until explicitly changed again to a different codeset. This allows a per-IE shift in interpretation of IE identifiers.

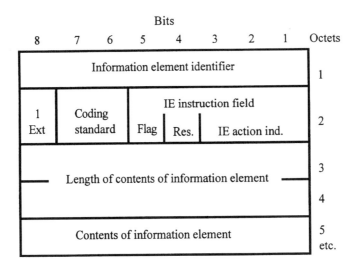

FIGURE 7.9 Q.2931 "generic" Information Element format. (From ITU-T Recommendation Q.2931.)

7.7 SUMMARY OF ATM SIGNALING

There are two categories of signaling within B-ISDN: metasignaling, which can act to set up a signaling channel, and out-of-band signaling (associated or non-associated). Out-of-band signaling occurs within the VPI on a specific VCI (5). The signaling entity makes use of primitives between itself, the higher layers, ATM layer, and management entity to allow a state table to be executed.

The state table consists of states, events, and actions. There are user-side and network-side states which are kept in parallel for the purpose of tracking the current state of a connection. Events are inter-layer primitives or the contents of messages (existing within the specific AAL_DATA_INDICATION primitive). Messages are made up of protocol discriminators, call reference values, and message types. A message type may contain zero or more Information Elements. The proper sending, and understanding of these messages and IEs will cause states to change and new primitives and messages to be sent.

7.8 SYSTEM NETWORK ARCHITECTURE GROUP (SNAG)

The System Network Architecture Group (SNAG) is a working group under the auspices of the ADSL Forum (but which works closely with the ATM Forum). SNAG is concerned with interoperability standards and architectures. Although they are primarily focused on use of ATM with ADSL as a long-term architecture, they also keep in mind the possibility that many sets of equipment may not want, or need, the full set of protocols that they propose.

Many documents contain a statement such as: "Protocol X over Protocol Y." This is from the OSI layering point of view. A Protocol X, residing in a higher OSI layer, will appear to be "over" the Protocol Y which occupies the lower OSI layer. However, from the protocol perspective, it might be better phrased to say that Protocol X is carried *within* Protocol Y. As we have seen in the discussion of ADSL frames and B-ISDN, data protocols will have a message (or messages) which allows for the sending or receiving of data. With this type of primitive, there will be a data field within which can be placed any type of data. Using the OSI layering principles, the type of data placed within the lower layer data field is unknown to the data carrying mechanism. In the other direction, the meaning of the contents is unknown but the layer must know what *type* of data is being carried in order to present it to the appropriate higher layer entity. It is also possible for the layer not to know about the contents or routing of the data—it can be handed to the C-plane functions to appropriately route the data.

There are two important papers that have been released by SNAG. The main "official" paper, Technical Report 10 (TR-010), is titled "Requirements & Reference Models for ADSL Access Networks: The `SNAG' Document." This document summarizes architectural requirements for interoperability. A longer "white paper" also exists called "An Interoperable End-to-end Broadband Service Architecture over ADSL Systems." In this paper, the following architecture is suggested.

ATM Over ADSL

There are two primary options for the architecture. In the first, ATM (using AAL type 5) is used over ADSL. The Point-to-Point Protocol is used over ATM and TCP/IP is used over PPP. Another possibility exists, however, which allows use of intranets without integrating ATM into the LAN. This uses a tunnelling protocol (such as Layer 2 Tunnelling Protocol, L2TP) and IP over a specific LAN protocol (such as Ethernet). This, in turn, can carry the PPP and TCP/IP elements needed for the "final" intranet use of the TCP/IP stack. Figure 7.10 shows a possible PPP over Ethernet Protocol Architecture based on L2TP, adapted from work done by SNAG under the auspices of the ADSL Forum.

The use of ATM over ADSL is a useful possible architecture. It may be unnecessary overhead, however, for many applications. In Chapter 8, we will discuss some other, simpler, alternatives that may find uses in the short-term and also, for certain applications, over the long-term.

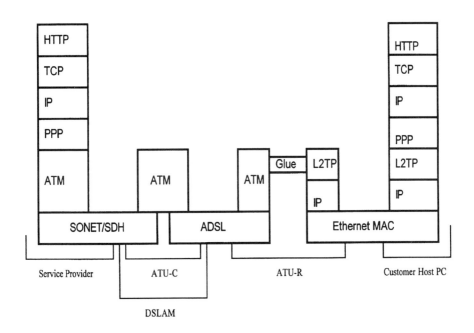

FIGURE 7.10 An example of tunneling use with ATM/ADSL and Ethernet.

8 Frame Relay, TCP/IP, and Proprietary Protocols

Making effective use of ADSL requires data transport protocols on top of it—although some types of "streaming data" such as video applications may contain their own implicit framing structure. Streaming data are video or audio which contain "self-extracting" data; e.g., data can indicate a certain frequency and amplitude. If bytes are lost in the stream, the effect will be a degraded voice, music, or video signal, but it should be understandable as complete data are not required.

In most cases, however, there will need to be higher-layer methods of data encapsulation to ensure that erroneous data is not acted upon. This may be a proprietary protocol (especially for services such as archival or retrieval services). A degree of control is needed for some services. For example, an ADSL video feed for "movies-for-pay" needs a way to send security, financial, and selection information. This, in turn, may trigger streaming data or a DVD-oriented MPEG-II set of data.

Proprietary protocols are unique within the particular application and manufacturer, therefore, only a discussion of possible protocol needs is directly relevant.

Two of the data protocols most often used within higher speed (or variable speed) networks are Frame Relay and Transmission Control Protocol/Internet Protocol (TCP/IP). TCP/IP is called a "protocol suite" because the two layers are normally found associated with each other (although, as we saw in Chapter 7, it's possible to use a tunnelling protocol with IP to provide intranet layering). Frame Relay was designed to provide a highly efficient transport mechanism. At its simplest, most efficient, level it gives only two services: data transport and status of the activity level of the circuit that is being used for transport.

8.1 FRAME RELAY

Frame Relay supports two types of circuits: Permanent Virtual Circuits (PVCs) and Switched Virtual Circuits (SVCs). Public network support of Frame Relay PVCs has been in effect since the early 1990s. Frame Relay SVCs are gaining importance in networks.

Frame Relay service architecture is similar to the architecture designed for ITU-T Recommendation X.31 which provides access to the X.25 networks via ISDN. The common architectural features are primarily associated with the way that circuits can be established. Both X.31 and Frame Relay architectures refer to two different cases. Case A uses out-of-band signaling in a generic form. This allows use of a circuit for non-differentiated data transport. In other words, the original switched network does not know the purpose for which the data channel has been initialized.

Case B provides integration of the data protocol and the signaling network. Thus, a user can request that a Frame Relay circuit be established in the out-of-band signaling. The user can specify particular parameters that should be associated with the circuit that is being set up. Thus, after completion of the out-of-band signaling, a data channel with specific parameters is available for use. This is called "on demand" signaling.

ITU-T Recommendation I.122 initially proposed a frame-mode bearer service (Frame Relay) in 1988. Recommendation Q.922 (based *very* closely on the V.120 data link layer, which is another data protocol used for terminal adaptation) defined the data link layer and Q.933 (later) defined the network layer for use in signaling applications. Q.933 is a direct variant of Q.931, giving an improper subset (reduced states and messages but with additional Information Elements not normally used in Q.931) of the N-ISDN signaling protocol. The states of Q.933 are similar to that discussed for Q.2931 in the previous section. However, the protocol discriminator is 8 (as is true for N-ISDN) and it can be multiplexed with Q.931 over the same data link if desired.

One important note is that Recommendations Q.933 and Q.922 provide a large set of options for the user. While options add flexibility, they also add complexity. Complexity reduces efficiency and can cause interworking problems. In conjunction with these, emphasis has been made to run the simplest set of protocols. The Frame Relay Forum, in their agreement documents, has specified particular subsets that the members have agreed to support. We will focus primarily on these subsets since they are more likely to be found as part of a network.

8.1.1 Frame Relay Data Link Layer

The data link layer for Frame Relay is defined by ITU-T Recommendation Q.922. This provides the definition for a protocol known as Link Access Protocol for Frame Relay (LAPF). LAPF is a variant of Q.921 (LAPD) and is very similar. Differences include an extended address field and the ability (Case A Q.933) to support additional signaling links on the D-channel for the Frame Relay circuits. A very important part of Q.922, however, was an "annex" called Annex A. This annex described the "data-link core" capabilities needed for fast transport of PVCs and this is what has been primarily used in the field.

Q.922A defined the data link core protocol needed for data link transport and identification. However, it did not have any information about the status of the link and, at that time, Q.933 had not been released. The ANSI T1 committee decided that Frame Relay support was particularly important in North America where N-ISDN was delayed in deployment because of tariff issues, some extra (debatably unnecessary) complexities of the initialization procedures, and a speed gap in comparison to the LAN speeds more widely in use in North America than in Europe.

Thus, ANSI released a series of recommendations numbered T1.606, T1.617, and T1.618 to satisfy initial manufacturing and equipment interworking requirements for Frame Relay. The SVC determination was probably adequate but, since it was

Frame Relay, TCP/IP, and Proprietary Protocols   131

not an agreed-upon international standard, many implementors waited until the issuance of ITU-T Q.933 before serious work on SVCs was done. However, Annex D of T1.617 provided an extension called the Local Management Interface (LMI). This allowed for STATUS_ENQUIRY and STATUS messages to be exchanged between user and network nodes on a Frame Relay network and completed the basic needs for connectivity for PVCs. Other companies, such as Stratacom and Sprint, were in the forefront of Frame Relay deployment and they offered some additions not present in the ANSI or ITU-T documents.

Q.922A provides only an unnumbered information field for transport of data. This means that it is not possible, at the data link layer, to determine that a frame has been "lost" during transmission, nor is there a method of retransmitting a frame once it has been determined that it has been lost. Frame Relay is one of the "new" protocols which have been created to be used in the context of modern data networks (with digital long-distance networks and many fiber-optic relays) which are *much* more error-free than the networks that existed during the time that protocols such as X.25 were created.

Generally speaking, unless a protocol doesn't *require* all data (such as a video or audio data stream), there will be a sequence number associated with a data group and some type of Frame Check Sequence (FCS) to ensure that the contents have not been corrupted. If the line is "noisy" and many errors and retransmissions are expected, it is best to perform error recovery at the lowest level practical. This prevents extra protocol processing (because every layer has a degree of processing required). However, if the line is relatively error-free, it makes sense *not* to do time- (and space)-consuming error recovery protocols that add extra overhead to *every* frame. Reduce the overhead for the protocol and make it a bit harder for higher layers to perform this *infrequent* error recovery. Frame Relay data link core protocol was designed with this criterion in mind.

8.1.2 LINK ACCESS PROTOCOL FOR FRAME RELAY

ITU-T Recommendation Q.922, as mentioned before, is a variant of the Q.921 set of data link layer protocols—most similar to the V.120 rate adaptation protocol. The frame structure has three primary fields: address field, control field, and (depending on the control field) an information field. Q.922, as is true of Q.921, is an HDLC protocol so it will be enclosed (or "framed") by flag characters (value hex 0x7E) and finished with a set of CRC bytes (CRC-16 is normal for LAPD and LAPF).

8.1.2.1 Address Field

The address field, as seen in Figure 8.1 can have multiple forms depending on the length of the address desired. The Frame Relay Forum (and Q.922A) documents indicate that *only* the default address field should be supported, but there are implementations that make use of the larger address formats. Generally, the larger address formats are likely to be useful only for network to network interfaces. The length of the address (and form of the address field) is determined by the 'E/A' bit—or Extension/Address bit—which is set to '1' for the last byte of the field.

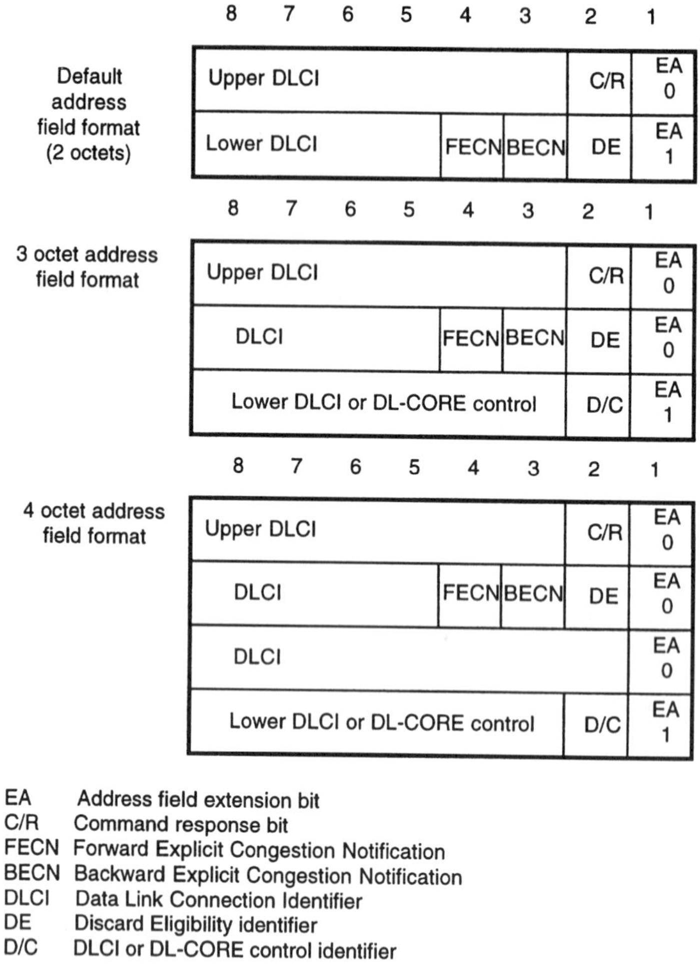

FIGURE 8.1

	8	7	6	5	4	3	2	1
Default address field format (2 octets)	Upper DLCI						C/R	EA 0
	Lower DLCI				FECN	BECN	DE	EA 1

	8	7	6	5	4	3	2	1
3 octet address field format	Upper DLCI						C/R	EA 0
	DLCI				FECN	BECN	DE	EA 0
	Lower DLCI or DL-CORE control							D/C	EA 1

	8	7	6	5	4	3	2	1
4 octet address field format	Upper DLCI						C/R	EA 0
	DLCI				FECN	BECN	DE	EA 0
	DLCI							EA 0
	Lower DLCI or DL-CORE control							D/C	EA 1

EA   Address field extension bit
C/R   Command response bit
FECN   Forward Explicit Congestion Notification
BECN   Backward Explicit Congestion Notification
DLCI   Data Link Connection Identifier
DE   Discard Eligibility identifier
D/C   DLCI or DL-CORE control identifier

FIGURE 8.1 Frame-mode address field formats. (From ITU-T Recommendation Q.922.)

The default address field has five fields. The Command/Response (C/R) bit has a more fixed structure than with Q.921, which is an "asymmetric" protocol. The value of '0' indicates a command frame and '1' a response frame. The Data Link Connection Identifier (DLCI) is split between the two bytes and forms a 10-bit address (possible 1024 addresses). The FECN, BECN, and DE bits are used in conjunction with congestion control. The D/C bit, in the three and four byte formats, controls whether it is the final part of the DLCI or the DL-CORE control identifier (which, for signaling, is the Unnumbered Information, UI, value of 3). Thus, we can see that if the final byte is really a DL-CORE UI control byte, the D/C bit will be set to '1' because the value of 3 sets the last two bits of the byte to '1'.

8.1.2.2 Congestion Control

A Frame Relay Network will make multiplexed use of all of the physical links within the network. This means that it will be possible for a specific Frame Relay channel to exceed the capacity of one of the links available. This traffic engineering is concerned with network congestion. A subscriber can, as part of Quality Of Service (QOS), specify a minimum transfer rate and latency time.

Congestion control is the process of deciding just what frames will be able to make use of the network and in what percentages. This will be determined by the subscribed QOS and the current status of all active links. For example, it is "reasonable" to allow a Frame Relay link to exceed minimum QOS parameters if no other links are presently using the physical link. However, if one link has been using their minimum (or greater) throughput usage for an extended period of time and other links are having problems being allocated their QOS throughput, then it is "fair" to give more bandwidth to links that have not been using the network as heavily.

Note, however, that a subscribed minimum should be expected to be met. This says that Frame Relay network providers should always provide enough bandwidth for the minimum throughput for *all* subscribers. This is rarely done, so traffic engineering allows for supporting the minimum throughput for only a subset. This is fine if the traffic statistics are such that each subscriber receives their minimum throughput, when used. If they do *not* receive the expected QOS, the network provider is not meeting contract conditions. Unfortunately, tools are only now starting to emerge in the market to verify that the QOS is being met.

Assume that minimum throughput QOS is being met by the network; nevertheless, the total amount of traffic is greater than can be supported. Congestion control then requires use of either congestion avoidance or congestion recovery mechanisms (or both). Congestion avoidance requires active help by the user equipment. Congestion recovery makes sure that the network doesn't collapse because of excessive use (and unforeseen buffer overflows and time-slice problems).

The Forward Explicit Congestion Notification (FECN) and Backward Explicit Congestion Notification (BECN) bits allow communication of network conditions to both user and network nodes. Since nothing can be enforced on the user equipment, specific algorithms are not generally recommended. However, if a user equipment receives frames with FECN set, it should start expecting that received frames will be lost soon. If it is able to do something (in the frames being sent towards the network) to reduce the traffic flow coming its direction then that would help the congestion. If it receives frames with BECN set, it is an indication that it should stop sending so many frames (if possible) or to set the Discard Eligibility (DE) bit on frames such that it has a certain degree of control over which frames the network will discard if it must reduce traffic flow.

The FECN, BECN, and DE bits should not be changed by the network nodes when they are forwarded from node to node. Thus, a node may set the FECN or BECN bits based on traffic conditions, but it should not clear them if they have been received from another network node. This is because the network nodes only have local knowledge of congestion conditions. If a user receives a BECN or FECN bit set to '1', then *some* node is having congestion problems, but it is not necessarily the network node to which it is directly connected. The DE bit should only be set by the consumer equipment. It is "cleared" only by the process of discarding the frame that contains the bit.

In the direction of the user equipment toward the network, only the BECN bit should ever be set by the user equipment. This is a notification that it is having difficulty handling the data traffic it is receiving. This would be an instance of the equipment being underengineered rather than the network.

ITU-T Recommendation I.370 defines two points in congestion recovery mechanisms. Point A indicates a degradation in service. Network buffers are being filled and there is an increase in the latency delay associated with data but minimum QOS can still be achieved. Point B is a degree of congestion which requires discarding data to keep the network functional.

8.1.2.3 Control Field

The control field of Q.922A is allowed the same message set as is Q.921. These include messages which allow entry into *multi-frame establishment* state. In this state, flow control is maintained, sequencing is checked, and retransmissions may take place. Q.922A does not make use of any control fields except for the Unnumbered Information (UI) control field. Therefore, we will not discuss other states and control fields, or other full Q.922 additional features in this book. Readers who want more information about full Q.922 should study ISDN or Frame Relay-specific books such as those mentioned in the references.

The user of only the UI field for data transmission reduces complexity for the data link core. No states are really necessary (although it must be true that the physical line is active and able to receive and transmit data) and data are passed through the data link core with only the addition of the address field (OR address and control field). Note that Q.922A and the FRF documents indicate that, for data passage, *no* control field is needed (including the UI control field). In order to be compliant with the standards, the firmware controlling the data core link should insert the UI control field only for locally (not passed from higher layers) generated data. This would include Q.933 signaling data passed within the protocol stack and the exchange of STATUS and STATUS_ENQUIRY frames. Insertion (and removal) of the UI control field in higher-layer data traffic may be an implementation option.

8.1.3 DATA LINK CORE PRIMITIVES

The interlayer primitives defined in ITU-T Recommendation Q.922 (Annex A) for the data link core primitives are shown in Table 8.1. These differ from the full LAPF primitives in that the DL_ primitives between the data link layer and network layer are not used (replaced by the DL_CORE_DATA primitives). This is true because multi-frame establishment state is not supported by the data link core. MDL_ERROR, MDL_UNIT_DATA, and MDL_XID are also not supported in Q.922A. Finally, PH_ACTIVATE and PH_DEACTIVATE messages are not supported. This is because Frame Relay PVCs are transported over "Case A" circuit-switched bearer channels or "semipermanent" TDM channels. In the first case,

Frame Relay, TCP/IP, and Proprietary Protocols

TABLE 8.1
Q.922A Data Link Core Primitive Types

	Type				Parameter		
Generic Name	RQ	IN	RS	CF	PI	MU	Message Unit Contents
Layer 3: Layer 2 Management							
M2N_ASSIGN	X	—	—	—	—	X	DL-CEI, DLCI
M2N_REMOVE	X	—	—	—	—	X	DLCI
DL-Core user: DL-Core							
DL_CORE_DATA	X	X	—	—	—	X	UI carried data
Layer 2: Layer 2 Management							
MDL_ASSIGN	X	—	—	—	—	X	DL_CORE CEI, DL-CEI
MDL_REMOVE	X	—	—	—	—	X	DL_CORE CEI
DL-Core: Layer 2 Management							
MC_ASSIGN	X	X	—	—	—	X	DLCI, DL_CORE CEI
MC_REMOVE	X	—	—	—	—	X	DLCI
Layer 2: Layer 1							
PH_DATA	X	X	—	—	X (BRI)	X	Data link layer peer-to-peer message

(RQ) Request; (IN) Indication; (RS) Response; (CF) Confirmation; (PI) Priority Indicator; (MU) Message Unit.

Source: From ITU-T Recommendation Q.922.

activation should occur at the time that the out-of-band signaling indicates that the channel is available. In the second case, activation should occur after power-up. Deactivation should occur only when out-of-band disconnection occurs or, as a transient condition, for semipermanent TDM channels. Thus, no explicit activation or deactivation commands are needed for the Data Link Core module.

The first item of note in the Data Link Core (DLC) primitives is that there are parallel _ASSIGN and _REMOVE primitives. These are for Layer 3 (management) to layer 2 management, Layer 2 to layer 2 management, and DL-CORE to layer 2 management. The purpose of these three separated (but very similar) sets of request (and one indication) primitives is to allow for a newer ITU-T architectural feature—layering within the management entity (or M-plane). It is thus possible to have a data link (Q.922) module, a data link core (Q.922A) module, layer 2 management and layer 3 management module. Although not directly mentioned in Q.922A, it is assumed that there will be a Layer 3 to layer 3 management set of primitives available also.

So, a possible primitive flow would be Layer 3 deciding (network-side must do this for PVCs) to assign a new DLCI link for use. It does a M2N_ASSIGN_REQUEST to layer 2 management. Layer 2 management then does an MC_ASSIGN_REQUEST to the DLC to allow use of the DLCI. Another flow would make use of the STATUS/STATUS_ENQUIRY messages. The terminal sends a STATUS_ENQUIRY message to the network which responds with a STATUS message. A new PVC is available, so the DLC sends an MC_ASSIGN_INDICATION primitive to layer 2 management. Note that, if Q.922 is not implemented or the Layer 2 and Layer 3 management modules are not split, only the MC_ primitives are required.

A minimal set of primitives for the DLC are MC_ASSIGN_REQUEST, MC_ASSIGN_INDICATION, MC_REMOVE_REQUEST (the assumption being that the management entity will handle the removal of DLCIs that are missing from STATUS responses). In addition, DL_CORE_DATA_REQUEST and DL_CORE_DATA_INDICATION are needed to transport data between the DLC and layer 2 or layer 3 and PH_DATA_REQUEST and PH_DATA_INDICATION pass data across the physical layer (or LLD, as discussed in Chapter 5) to DLC boundary.

An item of note is that the DLC primitive set does not support the MDL_XID primitive, which is necessary for Consolidated Link Layer Management (CLLM) messages. This message supports explicit notification of the causes of congestion or the current status of network resources.

8.1.4 Network Layer Signaling for Frame Relay

Signaling may take place once the connection to a Frame Relay network is in place. Although only Case A of Q.933 is presently supported by the FRF agreements, it does not really matter just how the connection has been made available for use. It is also possible (and likely, considering modern network conditions) to continue to make use of PVCs in parallel to SVCs. Therefore, it will be necessary to keep a common table of DLCIs in use to prevent overlap use of DLCIs between PVCs and SVCs. A Frame Relay link should be able to be distinguished by a combination of the DLCI and the TDM channel.

ITU-T Recommendation Q.933 is a very similar signaling protocol to that of Q.931. It also has a state protocol very similar to that of Q.2931, discussed in the previous chapter. These messages are sent over DLCI 0 within the Frame Relay TDM channel. (Stratacom, in their early Frame Relay implementations, made use of a different DLCI for signaling needs.)

Table 8.2 shows the Q.933 Frame-Mode Connection Control Messages. Frame Relay Forum agreement 4 (FRF.4) reduces the states and messages available in the support of SVCs. FRF.4 does not support ALERTING or PROGRESS. However, it does support the DISCONNECT message causing the state numbers (and transitions) to be somewhat different from that of Q.2931. Basically, states 11 and 12 are used for DISCONNECT messages (for receipt and sending) and state 19 is used when a RELEASE message has been sent. There is also no support of overlap sending or receiving, even in interworking applications.

FRF.4 also reduces the Information Elements (IEs) supported in the in-band SVC signaling. Only Case A is to be supported. Only two-byte address fields are supported. Only Q.922A is supported at the data link layer and so forth. The point of the reduction is to keep the protocol streamlined.

TABLE 8.2
Q.933 Frame Mode Connection Control Messages

Message	Reference
Call Establishment Messages:	
ALERTING	3.1.1
CALL PROCEEDING	3.1.2
CONNECT	3.1.3
CONNECT ACKNOWLEDGE	3.1.4
PROGRESS	3.1.6
SETUP	3.1.9
Call Clearing Messages:	
DISCONNECT	3.1.5
RELEASE	3.1.7
RELEASE COMPLETE	3.1.8
Miscellaneous Messages:	
STATUS	3.1.10
STATUS ENQUIRY	3.1.11

Source: From ITU-T Recommendation Q.933.

8.1.5 MULTI-PROTOCOL OVER FRAME RELAY

Since Frame Relay provides a streamlined, efficient mechanism for data transport of frames, it is well-suited for use as a long-distance transport mechanism for LAN protocols. The Internet Engineering Task Force (IETF) provides access to a large number of Request For Comments (RFCs) documents that define protocols to be used on, and with, the Internet. One of these is titled "Multi-Protocol Interconnect over Frame Relay" (RFC 1490).

The use of RFC 1490 is very simple. The idea is that a uniform structure of identification can be put on the beginning of any packets sent over Frame Relay. In this way, multiple protocols may be sent over Frame Relay and routed to appropriate applications.

This is really a Transport Layer (layer 4) item, although the actual location of insertion and removal of the RFC 1490 header is implementation dependent. A generic form of the RFC 1490 header is three fields. After the beginning flag (0x7E in hexadecimal) and the Q.922 address (assume two bytes) field, is where a control field (and data information field) would be followed by the 2-byte FCS and ending flag.

RFC 1490 says that the UI control byte will be inserted after the address bytes and, optionally, a padding byte (value 0) to bring the next byte to an two-byte address boundary, followed by the Network Layer Protocol ID (NLPID). If a 3-byte Q.922 address field was in use, the padding byte would never be needed since the three address bytes plus one control byte would bring the address to a two-byte boundary.

The NLPID values are administered by the ITU-T and the International Standards Organization (ISO). The NLPID may indicate a protocol such as IEEE Sub-Network Access Protocol (SNAP) which has its own identification mechanism. SNAP contains a three-byte Organizationally Unique Identifier (OUI) which precedes a two-byte Protocol Identifier. Thus, the sequence

$$0x03\ 0x00\ 0x80\ 0x00\ 0x00\ 0x00\ PID1\ PID2$$

will indicate a UI frame plus PAD byte for a SNAP header with an Ethertype indicated by PID1 and PID2. The following sequence

$$0x03\ 0x00\ 0xCC$$

will indicate a UI frame plus PAD byte with a NLPID indicating use of Internet Protocol within the Frame Relay Frame.

Thus, RFC 1490 is a simple chain of protocol identifiers. A UI control byte followed by a padding byte (which is an illegal NLPID byte to avoid confusion over optional padding bytes) and then an NLPID, which may indicate a type which has yet more hierarchical identification information.

8.2 INTERNET PROTOCOL

The Internet Protocol (IP) is probably the most highly advertised data protocol currently in use. This doesn't mean that it's the most widely used, the most versatile, or the most deployed. However, the Internet is probably the resource which is driving consumer DSLs to approach faster and faster access speeds. It is also probably the most important application to urge consumers towards the use of ADSL (or BRI ISDN, or whatever high-speed system makes the most economical sense for them).

The Internet Protocol takes approximately the same position as the Network Layer in the OSI Layer model. As such, it requires a data link layer underneath which may be as simple as HDLC framing with address information or may be ATM or Point-to-Point Protocol. "On top" of IP is often the Transmission Control Protocol (TCP) but, for tunnelling (or Intranet) applications, it is possible that it may be a tunnelling protocol such as Layer 2 Tunnelling Protocol (L2TP).

8.2.1 THE DATA LINK LAYER

The data link layer is concerned with making sure that the data has been transported to a specific address. The most ubiquitous protocol used with LANs is that of Ethernet, which will be discussed briefly in Chapter 9. Currently, the most popular protocol for use over the telephone network is Point-to-Point Protocol (PPP), although some older systems may still use the Serial Line IP (SLIP). We will not attempt to duplicate information better available in PPP books here.

The main criterion for the data link layer is that the IP frame is encapsulated so that the beginning, end, and length of the frame are known. Addressing information may also be required if the destination needs to know how to route the frame among multiple possible destinations.

8.2.2 IP DATAGRAMS

A datagram is a group of data that is "offered" to a system. Most IP traffic is concerned with such datagrams. The main aspect of a datagram is that there are no guarantees on arrival. In fact, part of the flow control, and congestion maintenance, involved with the Internet is basically the ability to discard frames when traffic becomes too heavy. Hopefully, the net effect of the algorithms used within the system will only result in a slowing of data and not an inability to communicate. However, recovery is left to the higher layers.

As is true for most protocols, the IP datagram consists of a header and a contents field (which may be empty). The header contains information concerning how the header should be parsed and how the data contents (if any) should be treated. Figure 8.2 shows the general structure of an IP datagram for Version 4. The minimum size for an IP header is 20 bytes which is broken down into five 32-bit words. The first 32-bit word has four fields. The version number occupies the first 4 bits (or nibble). This will often be the value of '4', but for IPv6, it will have a value of '6'. This information should always appear in a fixed location so the recipient can determine the form of the header. It could be located at some other fixed location, but having it at the beginning places fewer constraints on the structure of the rest of the frame.

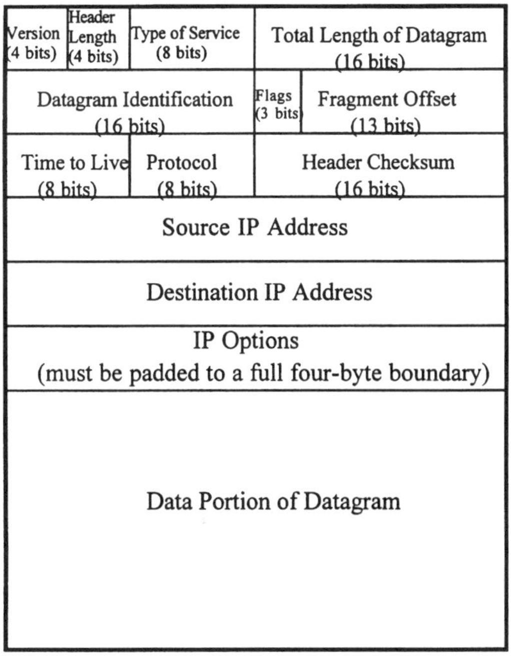

Version (4 bits)	Header Length (4 bits)	Type of Service (8 bits)	Total Length of Datagram (16 bits)	
Datagram Identification (16 bits)			Flags (3 bits)	Fragment Offset (13 bits)
Time to Live (8 bits)	Protocol (8 bits)		Header Checksum (16 bits)	
Source IP Address				
Destination IP Address				
IP Options (must be padded to a full four-byte boundary)				
Data Portion of Datagram				

FIGURE 8.2 Internet Protocol Version 4 header format.

The next nibble in the first word is the header length—defined as the number of *words* (32-bit groups) in the header. Since the values of 0 through 15 can be contained in a nibble, a maximum header size of 15 32-bit words (60 bytes) is possible. A value less than 4 is illegal. The next field is the Type of Service. The contents of this byte are set by higher layers. Three of the possible 8 bits are actively used in IPv4. These are bits used (by setting to '1') to indicate need for minimized delay, maximized throughput, and high reliability. The final field in the first word is the datagram length. This time, the length is in units of number of bytes. The 16-bit wide field thus allows a maximum of 65,535 bytes in the datagram *including the header*. Thus, a minimum-sized packet would have the value of 5 in the header-length field and 20 in the datagram-length field.

The next word has three fields. The first of these, the datagram identification field, is a unique 16-bit identifier assigned by the originator. Initially, there will be only one datagram with this identifier. However, in the process of routing the frame through possible other networks (some of which may be slower or have smaller frame limits) it is possible that the original datagram will be fragmented. The datagram identification will remain the same in each of the fragments. This will help in reforming the original datagram at the destination for interpretation.

The flag field in the second word has three bits available. In IPv4, only two bits are used. One indicates to the network "Don't Fragment" (DF bit). If the network is unable to do this, it will discard the frame and send back an error message. The other bit is a "More Fragments" (MF bit) indicator. Setting this bit to a '1' indicates that this is not the last (or only) part of the frame and that others will follow (the last making sure to have the MF bit set to '0').

The final field in the second word is the Fragment Offset. This indicates how many units from the beginning of the original datagram the contents of this fragment should be located. Since only 13 bits are available for this field and 16 bits are available for the datagram length, it is necessary to have the units mean 8 bytes each. This also means that each fragment should be a multiple of eight bytes so that the fragments can be put back together without gaps.

The third word has three fields: the "Time-to-Live," Protocol, and Header Checksum fields. Originally, the Time-to-Live field was meant to be the number of seconds the frame was allowed to be in the network before it would be deleted (unless received by the recipient and acted upon—at which time it would be deleted anyway). Thus, each network entry point would calculate the time needed to process and transmit the frame and decrement the field—if the frame didn't make it to the recipient before the field went to zero, it was deleted.

In practice, routing time is short enough that the time needed was often less than a second, it has become common for the Time-to-Live field to be treated as a "Number of Nodes" field and decremented for each node access. Another way of putting it would be to say that each node has a minimum decrement of one second (and that it is rare for the processing and transit time to exceed one second).

The protocol field indicates the type of higher layer (such as TCP) that should be interpreting the data. The Header Checksum is a 16-bit field that makes sure that the datagram header is not corrupted.

Frame Relay, TCP/IP, and Proprietary Protocols   141

The final two words (8 bytes) of the mandatory minimum 20-byte IP datagram header are the source and destination addresses. The 4-byte source address and the 4-byte destination address have their own network hierarchy and naming conventions which we will not go into for this book. Note that the IP Options field (which is not mandatory) is often omitted and rarely used.

We touched on IPv6 earlier. Although IPv4 is still predominantly used, Internet expansion will make a conversion to the newer standard necessary over time. IPv6 was first used as early as 1996 and was devised from feedback concerning real-world usage conditions. The main features of IPv6 are a greatly increased address space (16 bytes per address, or IPv4 to the fourth power) and a greatly reduced header size (decreased from 20 bytes to 8 bytes). The first nibble is still used to indicate version number and, therefore, how to interpret the frame.

Past the version field, there are five other fields that exist before the addresses. The Priority field replaces the Type-of-Service field and indicates something much more useful. In the mix of all frames coming *from the originating host,* what should be the priority of this frame. This allows the originating host to partially decide what messages (datagrams) should get highest priority whereas the form to Type-of-Service was much more node-dependent. The Flow Label field is primarily a place holder for a new feature called "flows" to be used (see RFC 2460). The Payload Length field is only the length of the data *following* the IPv6 header or headers.

The Next Header field indicates the type of header to follow the current header (IPv6 or subsequent). This field allows a cascading of headers prior to the datagram contents. Some of the additional headers (in order of how they should be presented, if used) are Hop-by-Hop Options, Destination Options, Routing, Fragment, Authentication, Enacapsulating Security Payload, and Destination Options headers. The final IPv6 header field (besides the addresses) is the "Hop" field which, as described earlier, is now officially the number of nodes through which the datagram has passed.

8.3 TRANSMISSION CONTROL PROTOCOL

The Transmission Control Protocol (TCP) is always used with IP, however, as we have seen, it is not always true in the other direction. IP may be used with other protocols such as the User Datagram Protocol (UDP) or L2TP. The fact that TCP is always used with IP, however, is why the combination is often used—TCP/IP. This combination (or any combination of layer protocols) can be called a "protocol suite."

TCP provides services very similar to that of Q.921 (although it is a higher layer), *but* with features associated with IP use and specifically expecting Internet types of applications. Applications using TCP expect the service to be reliable and efficient. Note that reliable does *not* mean infallible. However, it does mean that if a service (often meaning data transport) cannot be accomplished, the originator will be notified of the failure.

Note that some of the specifics of TCP will need to be modified to work in conjunction with IPv6—particularly the use of the larger address nodes. Also, certain value ranges may need to be increased, or shifted, to make better use of the larger address capabilities of IPv6.

8.3.1 TCP Virtual Circuits

TCP provides Virtual Circuits for the hosts to use in communication. These are neither SVCs nor PVCs as they are neither set up by signaling nor the same any time the connection is used. A TCP VC is determined by combining the IP address with a "port address" for both ends.

Port addresses fall into three categories. The first is "well-known ports." This is a way of saying that a port address has been allocated to a particular application. All numbers in this category must be approved by the Internet Assigned Numbers Authority (IANA) and will be in the range (at present) of 1 through 1,023. Port values in the range of 1,024 through 4,999 are considered to be "ephemeral ports." Ports higher than 5,000 are intended for use as specific, but non-official port addresses. For example, a company may have a special application for administrative or security use and will allocate a special address (say 5,551) for the port address associated with the application.

Ephemeral ports are essential for providing multiple VCs under TCP. A Virtual Circuit is composed of four components: originating IP address, originating port, destination IP address, and destination port. The destination port will probably be either a "well-known port" address or a "special specific" port address. There will most likely be only one port address for a particular application. This means that, for the duration of the TCP/IP connection for this service, three of the four components are fixed. Only the origination port address is available to allow multiple channels accessing the same application. Use of ephemeral port addresses thus allows for multiple channels to the same application.

8.3.2 TCP Header Fields

The fields of the TCP header are shown in Figure 8.3. The source port number is chosen by the originator from the possible ephemeral port numbers. The destination port number is either a "well-known port" or "special specific" port number. Next is a 4-byte sequence number, which is the identification number for the segment. In TCP, sequence numbers can start at any value and are encouraged to not start at 0 or 1 to avoid potential collisions during initialization of VCs. After transmitting a TCP segment, the internal sequence number counter is incremented. The Acknowledgment number is the number of the *next* expected received sequence number.

Say that Endpoint A sends a TCP segment with Sequence Number 17 and Acknowledgment number 534. If Endpoint B responds with a Sequence number of 534 and an Acknowledgment number of 18 then the sequence was successful. Both number fields are not always used, depending on the exact TCP sequence in progress.

The 4-bit header length is an indication of the size of the header in units of 4-byte words. Thus, the header can potentially be 60 bytes long. The next used field are the PCP flag bits which are used in negotiation and keeping track of the status of the connection.

The 6 bits of the TCP flag bit field are called: URG, ACK, PSH, RST, SYN, and FIN. The URG bit says that the segment is to be considered urgent and that the

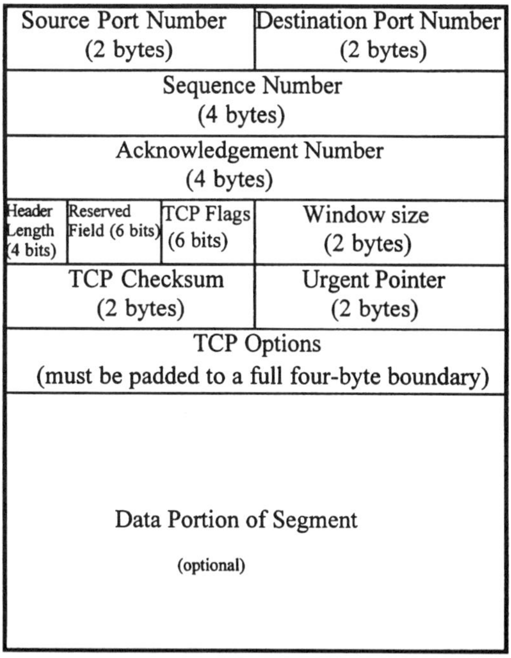

Source Port Number (2 bytes)	Destination Port Number (2 bytes)
Sequence Number (4 bytes)	
Acknowledgement Number (4 bytes)	
Header Length (4 bits) / Reserved Field (6 bits) / TCP Flags (6 bits)	Window size (2 bytes)
TCP Checksum (2 bytes)	Urgent Pointer (2 bytes)
TCP Options (must be padded to a full four-byte boundary)	
Data Portion of Segment (optional)	

FIGURE 8.3 Transmission Control Protocol header fields.

urgent pointer field should be used. The urgent pointer field is actually the offset of the last byte to be considered to be in this category.

The ACK bit is set after a segment has been received from the other end and should remain set in all subsequent messages to indicate that the acknowledgment number is valid. The PSH bit says that the TCP segment should be passed to the application even if the data length is not sufficient to warrant normal passage (used for short data such as a carriage return in Telnet applications). The RST bit requests the connection be reset when turned on.

The SYN bit is used during the initial synchronization handshake. It basically says to ignore any current sequence numbers that are stored for acknowledgment and to use the included sequence number as a new start for acknowledgments. The FIN bit indicates the finish of data—set in the last segment of a sequence of data.

The window size field is a dynamic value. It increases when the efficiency of the line is improving and decreases when latency delays are present. It is the number of segments that can be sent without having received acknowledgment. Thus, if the window size is four, segment numbers 113, 114, 115, and 116 can be sent *before* the acknowledgment number 114 is received back.

The TCP checksum is a 2-byte field that provides an error check of the frame contents, the header *and* a "pseudo-header" which includes the IP addresses. This provides not only a check on the data integrity, but also a check to make sure that the TCP segment has been given to the right IP destination address.

8.3.3 TCP FEATURES

TCP provides Virtual Connections allowing the same application to be used between hosts for different purposes. An example might be two different Telnet sessions running in two windows on a host. Sequence numbers allow for a check of missing information, as well as the possibility to receive out-of-order segments and still pass them to the application in the order they were originally sent. Windows help efficiency by allowing multiple segments to be passed before earlier ones are acknowledged—preserving the acknowledged nature of the data transfer while transmitting data even while waiting for acknowledgment of earlier segments. The TCP bits allow direct handshaking between two hosts to keep them synchronized.

8.4 PROPRIETARY PROTOCOL REQUIREMENTS

A short discussion on possible design requirements for proprietary protocols will finish off this chapter. The first category concerns lost data. Some forms of data allow loss of data without causing problems. This is usually applicable to "self-contained" data. Each datum provides useful information. Together all the data provide the "best" information, but perfection is not needed. This occurs in the analog speech network. It is also often true for digital audio and video signals.

The big question on network engineering is just "how good is good enough?" It is possible to lose 10 to 20% of the information in a speech signal without becoming unrecognizable. That doesn't mean that the clients will be happy about this and quality is certainly a valid marketing/pricing issue.

On the other hand, if you are sending a set of data that is totally dependent on the full data being present, then a loss of even one packet is unacceptable. Plus, the data must be reassembled in the correct order.

Preventing loss of data requires four things: a method to verify the data received is the same as the data sent (data integrity), a way to tell that the data received is the data expected (data identification), a way to tell the sender that the data has been correctly received (data acknowledgment), and a way to recover from loss of data (data recovery).

8.4.1 DATA INTEGRITY

Data integrity is usually performed by some type of checksum. A checksum occurs at the end of many items used in "everyday life"—the number for an "airbill" for tracking purposes, the number on a credit card, or the identification number associated with a credit or delivery account. While not all such situations include checksums, the larger companies will do such to *reduce* the amount of erroneous (or falsified) data received. A checksum is basically a set of numbers derived from the rest of the sequence. One popular method is to add all of the bytes together and then do a "1's complement" on the result. The following sequence

0x73 0x84 0xAC 0x11 0x23

would have the summing total of hexadecimal 0x0137. Assume that only a byte is used for a checksum (very generous for such a short sequence). This means that the value of 0x37 (or 0011 0111) would be the sum. The "1's complement" of such would be 11001000 or hexadecimal 0xC8. Thus, the sequence (with 1-byte checksum) would be:

$$0x73 \ 0x84 \ 0xAC \ 0x11 \ 0x23 \ 0xC8$$

The above example is a very simple checksum calculation. Note, for example, that the transposition of two bytes would cause the same sum (but would be an undetected error). More complicated algorithms are available, and used, including some that can (for relatively short sequences) indicate just *where* in the data sequence the error has occurred. The checksum is calculated and made a part of the data stream upon transmission. Upon receipt, the checksum is calculated again and if the checksums (sent and recalculated) match, the data is considered to have been received as transmitted.

8.4.2 Data Identification

If a data sequence has an identifier *and* the receiving entity knows what data identifier is expected, then the receiving side knows that the data was received in order and that no data was lost. This identification could be done by having each data frame identify itself and identify the next data frame. Normally, what is done is that the data identification will be sequential with "wrap-around." For example, if numbers 0 through 255 (using one byte) are used as identification, data frame 254 will follow data frame 253. After data frame 255, data frame 0 will be sent.

This can cause problems only if 255 packets are lost or if the sender (because of some error condition) starts renumbering its packets. Some protocols, such as Q.922A, will not use the "restart" value for identification, eliminating reinitialization of variables as a potential for confusion. As long as the identification length is shorter than the window size (see next section), no confusion is possible.

8.4.3 Data Acknowledgment

When the transmitter sends data (properly identified), it can either assume that they have arrived correctly or it can wait until the receiver sends back a frame indicating proper receipt. This "acknowledged" mode can slow transmission considerably, especially if the transmission distance is long. In order to be able to acknowledge data *and* not slow down the transmission capability, a window is often used. What this says is that a certain number of packets (the window size) can be transmitted before it must have acknowledgement (for the first or, in some protocols, for several frames).

As an example: endpoint A has a window size of 3. It transmits frames 5, 6, and 7 and then has to wait for acknowledgment before transmitting more. It then receives acknowledgment for 5 and can now send 8. (It might also receive acknowledgment for 6, implying that 5 also arrived, and then could send 8 and 9.) Generally, a system will attempt to be designed such that, under normal circumstances, no wait is needed. Thus, the transmitter would send 5 and 6 (and be able to send 7) and then receives acknowledgment for 5. If acknowledgment always occurs before the window capacity is reached, no delay of transmission is entailed.

8.4.4 Data Recovery

The above section on data acknowledgment assumed that no frames were lost. What happens if a frame *is* lost? Usually, the receiver will discard any frames that don't match what it expects. It can also do one of two possible notifications of error: implicit or explicit.

An implicit notification of loss is done by not sending an acknowledgment. This will work if the sender has a timer set at the time of transmission. If the timer expires without acknowledgment, the last unacknowledged frame is transmitted (plus, possibly, other frames that follow). Since the receiver discarded all other frames, the normal receipt/acknowledgment sequence can resume.

An explicit notification will have the receiver send a frame indicating an out-of-sequence frame (and, therefore, lost packets). This allows the transmitter to start retransmitting more quickly as the time delay would only be the amount of time needed to transmit the lost frames plus the time needed for one successful (but out-of-sequence) frame plus the time needed for the rejecting message. Explicit notifications make the protocol more complicated. Implicit notifications require timer customization. The choice depends largely on how often errors are expected to occur.

8.4.5 Data Protocol

Other than making sure data are transmitted correctly, a proprietary protocol may want to multiplex different information over the same logical link. This is done by identifying the type of data, or the logical entity for which the data is intended. "Control fields" are often used for this. A control field may indicate a frame used for initialization, idle checking of the link capacity, errors, or type of data. It is also possible that these types of frames will not all be legal at all times.

For example, it is not uncommon for data frames to not be legal until after an initialization sequence has occurred. Any situation where certain events are legal at certain times and others legal at other times is one where states are required. See Chapter 5 for state machine design.

We have now discussed data transmission using ADSL. Chapter 9 concerns data transfer at the remote unit from the protocol receiving unit to the host processor.

9 Host Access

Unless a data protocol device serves as a gateway (which can be the case for an ATU-C), there will be a need for efficient transmission between the host processor and the interface device. Generally, an access speed of twice the greatest aggregate data rate will be needed to allow for handshaking and overhead in the data transfer process.

Since ADSL is a technology that attempts to make use of the existing infrastructure (the local loops), it is not surprising that various methods have been discussed for the data transfer between an ADSL interface board and a host. These methods include an old "standard" interface called the Ethernet. The most prevalent form of Ethernet operates at 10 Mbps. Regular Ethernet speed is probably not fast enough for full ADSL bandwidths, but it is certainly fast enough for ADSL "lite." The newer "Fast Ethernet," which can operate at speeds of 100 Mbps, is fast enough even for VDSL use.

Host access methods can be split into two categories. The first, more immediate, method is to use a transfer protocol between the two devices. Ethernet is presently the most common method of LAN implementation—causing the frequent reference to it as the "ubiquitous Ethernet." The Universal Serial Bus (USB) architecture, which was developed by a consortium of major software and hardware vendors, is the latest (and fastest) attempt at a modern replacement of the old serial and parallel port interfaces. The USB is presently implemented to support a data speed rate of 12 Mbps—a slight improvement over the "regular" Ethernet implementation but difficult to directly compare because of protocol overheads.

The second category is mostly applicable to host-controlled systems, but can be used for other coprocessor systems as long as the host interface on the interface board is self-contained or an overlay control protocol is used in conjunction with the mechanism. This can be what is called the "ATM25" interface, which is the ATM physical layer standard making use of UTP for support of speeds up to 25.6 Mbps. Since it is, in itself, a physical layer specification, protocol data will normally be expected to be processed on the host. That is, the ADSL interface provides (and transmits) an ATM cell stream across the ATM25 physical layer. This is a physical layer bridging mechanism. Although the concept is intriguing and may very well be part of the general ADSL host-access solution, physical layer translation is basically a matter of taking the data contents (Layer 2 and above) from one physical medium and placing it on another physical medium.

Another variant on the second category is the aspect of motherboard support. This can either be a data bus which is directly tied into the motherboard chassis on a personal computer, or extensions of leads from the main microprocessor chip.

(The latter requires special design of the microprocessor.) The second category, whether it entails physical layer translation or direct microprocessor access to the data stream, is more of an "in the future"-type of interface as it requires additional new devices and interfaces. Part of the appeal of ADSL is to keep as much existing equipment as possible. Thus, the first category is likely to more popular and use of the Ethernet will provide the greatest possible existing client base.

9.1 ETHERNET

The Ethernet was devised as a solution to connecting different devices (primarily general-purpose computers, but peripherals were also allowed) within a short distance. This distance can actually extend to 4 kilometers but, in practice, this is rarely done. Although this is a form of telecommunication, since the network is achieved through the use of data transmitted from one station to another, it was conceived as a connectionless network. In other words, machines that were connected together were part of a complete unit.

The Ethernet has a maximum number of nodes set at 1,024, which is more than enough for most LANs. Plus, with a 10 Mbps speed parceled out over a possible 1,000+ machines, this allows only 10 kbps average data per machine. Naturally, such a speed would not really be possible over a 10 Mbps network, because overhead would reduce the possible capacity. However, Ethernet (as is true of most LANs) expects a very "bursty" form of data transfers. One machine may communicate with another at 5 Mbps for 30 seconds (allowing transfer of over 18,000,000 bytes of data) and then be silent. Normally, some machine on the network will be the dominant depository. Note also that the data will be accessible by every device on the LAN, but it is considered "good manners" to only make use of data intended for your own machine.

Actually, the capability of looking at the data of other nodes would require special Ethernet software (change in algorithm for receiving packets). If microprocessor support was used for adddress recognition, the change in software would be even more difficult. However, since the software has the responsibility of address checking, there are security considerations with the use of an Ethernet as it is always possible for some node to have an erroneous, or deliberately redesigned, address recognition algorithm.

Besides having the data available to every node on the network, Ethernet is particularly "fair" in its allocations since there are no priority levels for data. Note also that data is effectively simplex—going in one direction at a time. This is an inherent aspect of the design of the Ethernet loop—where data sent out on the transmit leads will return on the receive leads. Assume that there is a cable going into and out of your computer (in reality, there is likely to be a "T" connection spliced into a loop that goes from machine to machine). How do you pick a direction? You are really only able to send the data *out* and it will be sent to all devices on the loop (including your own machine). Thus, the data is not simplex as much as it is a situation that only one machine can talk at a time.

Host Access

Ethernet has a very simple architecture. There is the physical medium which may be coaxial cable, twisted pair (including UTP, although faster access requires better quality twisted pair), fiber optics, or something else. The transceiver is capable of receiving and transmitting over the chosen physical medium. Repeaters may also be needed if the cable starts exceeding architectural limits. Finally, the transceiver is connected to an Ethernet controller card, which is able to talk to the transceiver and provides part of the Medium Access Control (MAC) which maps into the lower half of OSI layer 2.

9.1.1 History

Ethernet began in the early 1970s via experiments by Xerox, which wanted to connect together office equipment. It worked with Digital Equipment Corporation (DEC, now owned by Cabletron) and Intel to publish a standard for the use of this technology in 1980. This was Ethernet Version 1.0. The standard was further developed into Version 2.0 (which, unfortunately, was not fully compatible with the previous version). Possibly because it was obviously still an evolving standard, Ethernet was not developed by many companies at that time—primarily DEC and Xerox. However, the technology was proven to be architecturally feasible though still slower and more cumbersome than later implementations.

The Institute of Electrical and Electronics Engineers (IEEE) took over the specifications for Ethernet and created their own standard called ANSI/IEEE 802.3 (also adopted in similar form by ISO/DIS 88302-3). One advantage of IEEE taking over the specification was that it directly coordinated the Ethernet architecture with the OSI model. There were some changes from Ethernet v2.0, but these were mostly associated with the connector technology and the type fields so that older equipment could interwork with the new 802.3 frames with little change.

Of interest concerning the IEEE 802.3 standard (normally just referred to as Ethernet) is the nomenclature for the physical medium. This is referred to by three parts. The first part specifies maximum speed in units of 1 Mbps. The second part indicates the modulation type and the third part indicates the maximum length for each segment. Thus a 10base2 is a baseband technology limited to 200-meter segment lengths and a maximum data speed of 10 Mbps.

9.1.2 OSI Model Layer Equivalents

Figure 9.1 shows the parts of an Ethernet and how they correspond to the OSI model. The physical layer is split between the transceiver and the controller card. The Medium Access Control (MAC) acts as the bottom sublayer of the data link layer. Note that this diagram does not show layers 4 (transport) through 7 (application) at all. This is because they are not considered to be part of the Ethernet specification and are defined by their respective application specifications requirements. Actually, the Ethernet is only defined through the MAC. The LLC is expected to be protocol dependent. This is very similar to the division of the HDLC physical intensive aspects into the "upper" section (TC) of the physical layer.

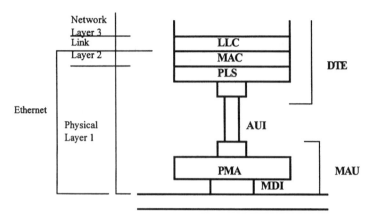

LLC   Logical Link Control
MAC   Medium Access Control
PLS   Physical Layer Signaling
AUI   Attachment Unit Interface
PMA   Physical Medium Attachment
MDI   Medium Dependent Interface
DTE   Data Terminal Equipment
MAU   Medium Attachment Unit

FIGURE 9.1   Ethernet sublayers and the OSI model.

9.1.3 THE MEDIUM ACCESS CONTROL (MAC)

The MAC, which is considered to be the lower layer of the layer 2, is responsible for putting data into frames and decoding the address fields. Thus, in an Ethernet, all of the frames must be read unless there is hardware support for the address recognition (not available in all hardware chip sets).

The transmission and reception algorithms are based on a method called Carrier Sense Multiple Access/Collision Detection (CSMA/CD). Therefore, this amounts to transmission by shouting into the wind and, while shouting, listening to see if anyone else is talking. If someone else is talking at the same time, there is a collision. Since both of your messages have already been corrupted due to the collision, you both stop. If you both waited the same amount of time to try again, there would never be any success as collision would occur every time.

Thus, the CSMA/CD algorithm has a random factor built into it that causes the transmitter to wait a random amount of time before trying again. Figure 9.2 gives a short transmission algorithm as described by the Ethernet specification. Obviously, this method results in considerable unavailable time on the network. The percentage will depend on the number of active nodes (nodes attempting transmission) and the average length of each transmission. It would be a good achievement to be able to use half of the potential transmission speed.

Other networks solve this problem by sending a short frame (called a "token") indicating that it wants to send data. In a true token network (possibly a token ring

Host Access

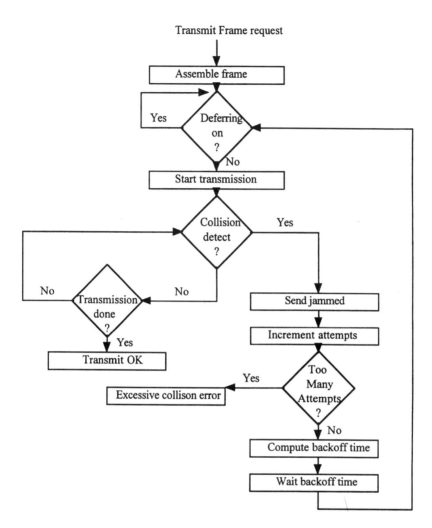

FIGURE 9.2 Flow chart for Ethernet transmission.

network, depending on the configuration of the network) there is a supervisor node with the right to allocate the right to transmit. This would not work on an Ethernet because all nodes have equal priority and equal rights to transmit. The best that could be done would be to have a cooperative situation where, when a "can I transmit frame?" was received, all other nodes waited a period of time (about 150% of the time needed to transmit a maximum-sized frame would be about right). Unfortunately, a collision on the "token request" would make the problem even worse.

Thus, Ethernet has low overhead and a very equalitarian and simple procedure. However, it has great potential for collisions and, thus, is not very efficient at being able to use the full bandwidth. The only time that the full bandwidth can be used is when only one node is active.

9.1.4 THE ETHERNET FRAME

The MAC frame for Ethernet has six fields: the preamble, destination address, source address, type/length field, data (plus optional padding bytes), and Frame Check Sequence (see Figure 9.3).

The preamble acts very similar to that of a flag in HDLC protocols. However, with HDLC, the beginning of a frame is detected by finding the first byte that is *not* a flag (byte value 0x7E). This works because the carrier is always "on" and frames are synchronized at the byte level. With the Ethernet, it must be assumed that each frame can start at any bit point in the transmission medium and that the medium is carrierless when idle.

Thus, a pattern is needed that can be recognized, and allow bit and byte synchronization and detection of the first byte of the frame. Ethernet does this with a preamble. The preamble has a pattern of 64 bits. The first 56 bits are a succession of '10' repeated 28 times. This allows the receiver to synchronize at the bit level—but does not allow byte synchronization. Byte synchronization is achieved by the final 8 bits of the preamble which has the "last" bit set to '1' rather than '0'. These last 8 bits are called the Starting Frame Delimiter (SFD). The preamble is shown in Figure 9.4.

The preamble is generated, and used, by the physical layer so it is not precisely the MAC layer (or part of the frame). However, like the HDLC flag, it is an integral part of the frame by marking the ends of the frame.

The address fields, destination and source, are six bytes each. They are further divided into two fields—a manufacturer identification and a Network Interface Card

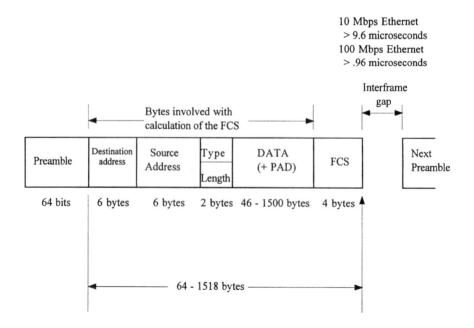

FIGURE 9.3 An Ethernet frame structure.

| 10101010 | 10101010 | 10101010 | 10101010 | 10101010 | 10101010 | 10101011 SPD |

FIGURE 9.4 The Ethernet preamble.

(NIC) identification. The manufacturer ID is allocated by a central authority—the IEEE.

IEEE 802.3 places a couple of other restrictions on the address. First, recognize that bytes are transmitted lower order bit first. Thus, the value of hexadecimal 0xCD (or 11001101) is actually transmitted as

$$1 0 1 1 0 0 1 1$$

One of the restrictions concerns the assignment of broadcast and multicast addresses. A broadcast destination address is all '1's (or a value of 0xFF 0xFF 0xFF 0xFF 0xFF 0xFF). If the first bit (low-order bit) is a '1' then it is a *multicast* address meant to be sent to all of a group of machines (that are built by the same manufacturer). Another way of saying this is that the specific manufacturer ID always starts with an even number (least significant bit is not 1) for the first byte of the three-byte field. An odd byte for the first byte of the manufacturer ID will indicate a grouping of manufacturer's machines.

The next two-byte field is the type/length field. In Ethernet v2.0, this field was used to indicate the type of layer 3 protocol used in the transportation of the message. Within the IEEE 802.3 standard, this field was changed to a length field. There is still a desire to have Ethernet v2.0 and IEEE 803.2 frames work together (although the number of pieces of equipment that are implemented according to Ethernet v2.0 must be dwindling). This means that there must be some way of determining the meaning of the field. This is done because of limits on the contents when used as a length. Ethernet frames can be no larger than 1,500 bytes. This means that any value greater than 1500 (hexadecimal 0x5DC) can be used for type without conflicting in purpose.

What about the length? If the field is *not* used for length, then how can it be determined? The length can be determined by counting the number of bytes received between the last byte of the preamble and the dropping of the carrier (idle line conditions). The length is still potentially useful in cross-checking the frame (what if someone else started transmitting just after the last byte of a FCS?).

The data field must be at least 46 bytes long and no more than 1,500 bytes. The meaning of the contents is unknown to the MAC sublayer. If the data field is meant to be fewer than 46 bytes long, the rest of the 46 bytes must be filled in by some "meaningless" padding bytes. This points out another use for the length field in the IEEE 803.2 Standard—allowing the length field to be used for the *useful* data and allowing the PAD bytes and FCS to be delimited by the drop of the carrier.

The final four bytes of the frame are the Frame Check Sequence. Using a 32-bit Cyclical Redundancy Check, it will cover the bytes of the frame from the beginning of the destination address to the end of the data (or PAD bytes). The preamble is not included. The FCS/CRC is transmitted most significant bit first (per byte) just to keep the hardware designer alert.

9.1.5 Physical Medium and Protocols

One of Ethernet's greatest advantages is the great versatility that it has in accepting different physical media for transmission and reception. In part, this is because the protocol was defined at the bit level (with the preamble sequence allowing for bit and byte alignment). Note that, without use of the length field, there must be some type of ending frame delimiter (such as the carrier drop).

The original physical medium defined for use with Ethernet was a 50-Ohm coaxial cable. Other characteristics were also defined for use of the medium such as total resistance, attenuation levels, distance limitations, and so forth. The IEEE issues additional specifications by appending letters to the specification. Thus, 802.3a lists a definition of the use of 10base2, or "Thinnet" and IEEE 802.3b defines the use of 10broad36. Other standards exist for 10baseT (use of UTP or STP for the physical medium).

9.1.6 MAC Bridges

We have already noted that repeaters are part of an Ethernet LAN. A bridge links two Ethernets together that do not use the exact same physical medium (but they are still Ethernet LANs). The diagram in Figure 9.5 gives an example of an Ethernet to StarLAN bridge. Originally, the bridges were designed without much of a standard, but the IEEE issued 802.1 in 1990 and this caused some minor changes to the hardware.

The bridge can act as a router if two or more networks are connected together by using routing tables based on the destination address. It can also prevent certain nodes from communicating with other nodes on other networks. The MAC bridge differs from a repeater in that it will examine the incoming message through the MAC layer and then take the data from that frame and send it back down the appropriate link unchanged. It also acts as a special repeater in that it will eliminate

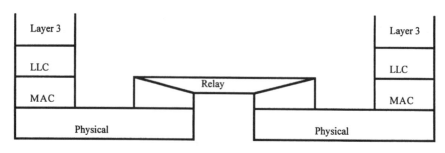

FIGURE 9.5 A MAC bridge between StarLAN and Ethernet.

any illegal frames (too short, too long, bad CRCs) because the Ethernet frame must be legal.

9.2 UNIVERSAL SERIAL BUS

The Universal Serial Bus (USB) was developed as a very local high-speed access mechanism. It was created out of a design consortium with Compaq, DEC, IBM PC Co., Intel, Microsoft, NEC, and Northern Telecom. The design is an open one (available at the http://www.usb.org website) and is royalty free. There are (currently) two related standards (accessible from the USB website) for the Open Host Controller Interface (OpenHCI or OHCI) by Compaq and the Universal Host Controller Interface (UHCI) by Intel.

The USB is a host-client architecture. Normally, the general purpose computer will act as the host and any peripherals will be the clients. Clients may use the USB in an upstream fashion (for peripherals like keyboards or mice) or in a downstream fashion (for monitors), or in a duplex mode (such as disk drives). However, the host will provide the general control of the peripheral device.

9.2.1 GOALS OF THE USB

The goals for the design of the USB are listed in Table 9.1. Basically, the goals were to have something available quickly that would allow high-speed transfers, be easy to configure, and allow expansion into new support areas without having to do a new design.

One of the USB's main goals was to provide "plug-and-play" operation for peripherals on personal computers. Thus, when a device is unplugged, the host OS would recognize that fact and refuse to route data to the (now missing) peripheral. When a device is plugged in, the host OS will automatically load an appropriate device driver that allows the host to communicate with the peripheral. It is also possible for hosts to be connected together by use of a special self-contained unit.

TABLE 9.1
Universal Serial Bus Goals

- Ease-of-use for PC peripheral expansion
- Low-cost solution that supports transfer rates up to 12 Mbps
- Full support for real-time data for voice, audio, and compressed video (stream data)
- Protocol flexibility for mixed-mode isochronous data transfers and asynchronous messaging
- Integration in commodity device technology
- Comprehension of various PC configurations and form factors
- Provision of a standard interface capable of quick diffusion into product (quick to implement)
- Enablement of new classes of devices that augment the PCs capability (growable)

9.2.2 USB Architecture

A USB system is split into three definitional areas. These are the USB interconnect, the USB host and the USB devices. The USB makes use of a tree topology as seen in Figure 9.6. The host is always in control of all the clients but some of the clients can be hubs which connect to other hubs of clients. In USB terminology, the distance from the host is considered to be a "tier." A tier 1 node is connected directly to the host while a tier 2 node is connected to a hub that is connected to the host, and so forth. Note that a hub can also be a device. A keyboard, for example, can be connected to the host and then a mouse connected to the keyboard (a similar configuration to that traditionally used with an Apple Computer's Macintosh™ and now used with USB on the iMac™).

The USB is a polled bus with the Host Controller initiating all data transfers. The host, on a scheduled basis, sends each connected USB device a "token packet." The device receives and parses the packet. The transmitter of data then sends the data and the receiver acknowledges receipt of the data.

The data stream, itself, may either be an unformatted (as far as the USB is concerned) bit stream or a USB defined message structure. Each data stream is called a "pipe." One pipe is created upon initialization—the Default Control Pipe.

Each pipe can be used for an upstream (from device to host), downstream (from host to device), or bidirectional data transfer. There are four types of data transfers. The Control Transfer allows configuration information to be passed and can allow control of one USB device by another (always through the host, however). Bulk Data Transfers are concerned with large, and likely short duration, groups of data being transferred. Interrupt Data Transfers are associated with events that need

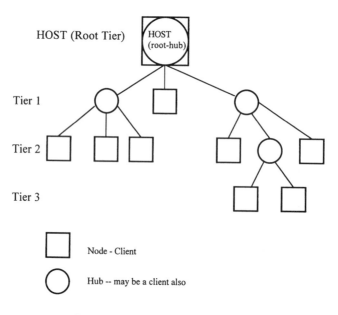

FIGURE 9.6 USB tree bus topology.

Host Access   157

feedback quickly (such as moving a mouse or typing a key on a keyboard). Isochronous Data Transfers occur when a pipe is allocated a percentage of the bandwidth for an ongoing data transfer (to be scheduled by the host).

9.3 MOTHERBOARD SUPPORT

The USB is connected into the motherboard although whether it is to be considered a data bus extension or microprocessor direct access depends on use of definitions. The Intel UHCI specification is primarily concerned with semiconductor device interfaces to handle the host side of a USB connection. It could "easily" be integrated into the microprocessor's circuits or it could remain a separate circuit.

On a more general basis, motherboard support is used whenever there is direct access by the general purpose computer's microprocessor to the device. Ethernet normally is *not* considered to be supported directly by the motherboard because the microprocessor will interact with the Ethernet controller board using a data protocol other than Ethernet.

In general, if the microprocessor interacts directly with the peripheral device, it is a microprocessor direct access. If it interacts via a separate circuit board but has direct access to the memory and functions of that board, it is a data bus extension. If it communicates with a protocol peripheral via a different protocol than what is being used by the communications device then it does not have motherboard support. (Note that whatever protocol is used to communicate with the peripheral [e.g., serial RS-232C] is supported by the motherboard.)

9.3.1 DATA BUS EXTENSION

The data bus extension is sometimes also called the machine bus or backplane. Some of the most common backplane architectures are known by the hardware and protocol standards involved with their use. These include the Industry Standard Architecture (ISA), Extended ISA (EIASA), Micro Channel Architecture (MCA), Peripheral Component Interconnect (PCI), Small Computer Systems Interface (SCSI), Versa Module Eurocard (VME), and Apple Desktop Bus (ADB). Circuit cards are designed to work with one of these backplane architectures. Note that some of these are slowly being replaced with newer architectures, that serve the same purpose, such as USB.

Circuit boards that make use of data bus extensions must be in a "powered-down" state to be removed or inserted. This is because there are direct links to the power and data leads of the microprocessor and other circuit boards and if the board is not removed exactly "right" it can cause damage to the rest of the system (as well as the circuit card being removed or inserted). This is one of the advantages to the USB. Since the power lead is isolated by the USB circuitry, it is possible to disconnect (or add) a device to the USB connection tree without powering down the device. Note that it *is* possible to design data bus extension (backplane) architectures for "hot" insertion and removal but it is more expensive and, if you have to remove the cover and work inside of the computer anyway, it is a generally good idea to have the power off.

9.3.2 Microprocessor Direct Access

This term means that there are pin-outs from (and to) the microprocessor that are directly involved with a communications protocol. Note that this *could* be a higher-level protocol such as ATM or the AAL but is likely to remain at the physical layer or MAC layer to maintain the possibility of multiple uses for the leads and microprocessor circuitry.

The manufacturer of a microprocessor will integrate a specific physical protocol support only when there is enough of a market available to be willing to pay back the development and manufacturing cost for the advanced chip. If enough manufacturers add a feature, it becomes "standard" and it is unusual to find a chip without the feature. If SNAG, or some other manufacturing or consumer group, succeeds in standardizing the protocol stack associated with ADSL use then it is open to have direct microprocessor support at any standard level.

In the next, and final, chapter we will talk about issues involved with software architecture with emphasis on ADSL. A traditional peek will be made into the future to try to see what types of advances are being undertaken and where they may lead.

10 Architectural Issues and Other Concerns

ADSL offers the possibility of relatively high-speed connectivity to a location that has twisted-pair wiring from a central location (central office) to the premises. However, being relatively new (particularly in deployment), there are quite a few issues that have to be considered before ADSL can take its place in the assortment of telecommunications protocols and devices that are generally used.

Some of the issues are associated with how the software needs to work with the various multi-protocol situations that are proposed for ADSL architectures. These are not issues from the point of making a large difference in the acceptance of ADSL. Rather, they are issues associated with being able to produce products that are able to be easily changed in an environment that is still volatile in regards to standardization.

Signaling issues still remain. In part, this is also a choice of just what services are needed for specific applications. Also, it is concerned with the speed of deployment of new, enlarged, high-speed data networks. There must be a shift from the existing speech long-distance network to one that is capable of transferring various data rates—from 1,200 bps for small e-mail messages to Megabit rates for live video feeds and mass data transfers. If older signaling protocols are used, it will continue to tie user equipment to the existing speech infrastructure. If newer signaling protocols are used, there is the danger of negative feedback if the demand outpaces the ability of the new network to grow.

In regards to this, standardization is very useful. ADSL provides a basic physical capability. However, there is nothing which uses the physical layer without additional protocols on top of it. It is possible to have many different devices making use of ADSL, just like MODEMs, telephones, and faxes make equal use of the existing Public Switched Telephone Network (PSTN). However, if the equipment of the ATU-R and ATU-C are not sufficiently standardized, there will be problems every time a user wants to connect different equipment.

ADSL cannot be used effectively without the entire system being able to support the speeds involved. Aside from general bottleneck situations (which occur today within the analog and BRI ISDN access world), more powerful PCs are required and high-speed connections to the host PCs to make full use of the potential of ADSL.

Migration needs will continue for the indefinite future. Migration to ADSL is only the first step to the access to the speeds envisioned for a global communications network. However, it must also be kept in mind that everyone does *not* need the same types of services and access. Thus, the shift of the infrastructure must be done in accordance with continued support of existing equipment and services.

10.1 MULTI-PROTOCOL STACKS

We have discussed a number of different protocols in conjunction with the use of ADSL. Since ADSL is a physical layer protocol, it is possible to just replace the "normal" physical layer with ADSL, given the appropriate standards to indicate how the MAC (or TC, in ATM terminology) sublayer is to be inserted into the physical layer. However, ATM is also primarily a data transfer protocol; it is not oriented toward particular applications. This means that other protocol stacks are likely to be used on top of ADSL or ATM over ADSL.

In this book, we have discussed the signaling (and alluded to data) protocols associated with BRI and PRI ISDN, Frame Relay, TCP/IP, and have used PPP and X.25 in examples. There are also proprietary protocols which, if ADSL equipment is ever standardized, may need to exist on top of (or within) the chained layer protocol stack associated with the standard. Even before potential standardization, it is likely that there will be older protocols used on top of newer protocols better suited to the high-speed needs of ADSL.

10.1.1 ARCHITECTURAL CHOICES

The first issue associated with multi-protocol stacks is that of choosing the appropriate combination of stacks for the application. The SNAG model gives us one possible configuration for the basic ADSL access (although it is more oriented towards Internet access than a general architecture would be).

Frame Relay over ADSL makes some sense because Frame Relay is not specifically associated with a particular bandwidth. For Frame Relay, however, we are faced with replacement of the physical layer (part of the physical layer, as mentioned before)—The HDLC components *without* the flags. In other words, since the medium has changed (or, at least, the physical use of the medium) the physical layer framing method must change. The HDLC frame from address field through the CRC would be put into an ADSL payload (possibly fragmented over several payloads). But, Frame Relay will often be carrying other protocols. The existence of the Network Layer Protocol IDentifier (NLPID) shows how important this is to general use.

The Internet is a driving force behind the attempt to upgrade the infrastructure and make use of new technology to have faster access. This means that TCP/IP is very important. TCP/IP can also be carried directly over ADSL although it is more likely that synchronous TCP/IP frames will be designed to be carried within PPP frames to provide greater link security and flexibility. PPP is an HDLC-based protocol (though not related to the Q.921/LAPB/Q.922 set of protocols) so use of it over ADSL is very similar to that of Frame Relay.

Since there is quite a bit of BRI ISDN equipment deployed, it is very possible for it to be sent over ATM over ADSL. It is not likely to be sent directly over ADSL, since the ADSL standards (particularly the international standards) already allow

Architectural Issues and Other Concerns   161

use of the BRI ISDN occupying the low baseband area underneath the ADSL spectrum.

If ATM over ADSL becomes the general standard for use of ADSL "modems," then there will be quite a bit of protocol stacks set up. Normally, AAL Type 5 will be used for this because it allows any type of information to be conveyed with the least overhead (assuming that the protocol being encapsulated will have its own various error checks and retransmission capabilities).

10.1.2 Software Implementation

There are various ways to implement a protocol stack. Many of the design criteria depend on just how many logical links are to be supported and whether it is for a single protocol or is to be able to support multiple protocols (not at the same time, unless they are being supported over different logical links).

A simple, single protocol stack can be looked at as a string of functions. Layer 4 calls layer 3 which calls layer 2 which calls the physical layer for transmission. On the receive side, the LLD sends the data to the data link layer which calls layer 3 that calls layer 4, and so forth. Note that, in accordance to our discussion about LLDs in Chapter 5, there is a "software break" between a received frame in the LLD and the processing of it in the data link layer. For fast data protocols, this is almost a requirement in any type of protocol stack.

The software break can be done in many ways. Use of an operating system, or real-time tasking system, will allow direct function calls to support the message system. The same can be done for simple systems by use of a "mailbox" and a semaphore. The semaphore is checked and, if available, is set and vital information associated with the primitive ia put into the mailbox. The LLD continues on with its task of polling, or being interrupted by, physical events. The semaphore is cleared by the entity servicing (or reacting to) the primitive put into the mailbox.

If the semaphore is already set, it is necessary to wait until it is cleared. If interrupts are disabled (as they would be in an interrupt service routine) then it is quite possible that the software which *would* clear the semaphore will never get a chance to do so. Thus, a simple semaphore/mailbox method will only be satisfactory if the ISR is the only one that might set the semaphore.

A more general algorithm, therefore, is the minimum amount of a software break that can be inserted between the physical LLD and the upper layers. This requires a queue of mailboxes to be available so that one is always free. A set of semaphores (one per mailbox) is also possible to use but this requires coordination of just *which* mailbox is to be used and serviced first. It is better to have a single semaphore but a queue of mailboxes.

The algorithm for this would be as follows. Check the semaphore (it may be set but, if so, it will only be set for the time needed to process the mailbox queue and *not* the time needed to process the data). As soon as possible, set the semaphore to lock out others from manipulating the mailboxes. Find the next mailbox to be used. Set the "used" field in the mailbox and fill in the primitive information. Release the semaphore. This method doesn't eliminate lock-out periods, but reduces them to fixed lengths of time (time needed to find an empty mailbox and fill it with primitive information) that are acceptable.

10.1.2.1 "Physical Layer" Replacement

One of the easiest ways to add protocol stacks, that are implemented according to the OSI model, is to take the "innermost" stack, remove the bottom layer (physical layer) and replace it with a "glue" layer that is able to interface with the network layer of the next "outer" stack. Figure 10.1 shows an implementation of this.

The glue layer will deal with physical layer primitives from the upper layer and network layer primitives towards the network layer of the next stack. So, Frame Relay over ATM would replace the use of an HDLC LLD driver with access to the AAL.

This only works if the layers are "pure." Therefore, the data link layer of the Frame Relay stack in this example would have to be responsible for all of the HDLC formatting and parsing except for the physical layer idle character and zero-bit insertion. It is possible to use a pseudo-data link layer that does not insert the CRC if it is known that, upon receipt, the data will be fed into the stack at the same spot (i.e., after the CRC has been inspected and removed). This also implies that the other stack that is being used must have the capability of error detection.

This is another way of indicating that a protocol is to be encapsulated within another. The difference is only in showing how this is to be done. An HDLC frame is composed of an address field, control field, optional data bytes, and an FCS/CRC field. The purpose of the CRC is to make certain that the bytes received have not been corrupted. If the data protocol within which it is encapsulated is also making sure that the bytes have not been corrupted, there is no need of the CRC. However,

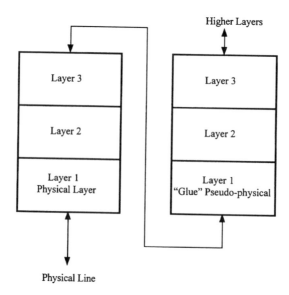

FIGURE 10.1 Protocol stacking at the physical layer.

Architectural Issues and Other Concerns

if upon receipt the CRC is expected to be present, the transfer will fail. For maximum compatibility, the full data link layer frame should be sent so that further "bridging" keeps the frame intact as needed by the protocol.

10.1.2.2 Coordination Tasks

We have said that there must be "glue logic" (very similar to that connecting disparate semiconductor circuits together) allowing use of one protocol stack's higher layer in place of the lower layer. However, in the case of non-permanent circuits, this will involve stages of readiness.

The "outside" protocol must be ready before the next "inside" protocol is able to be sent. This is true within other OSI model protocol stacks but the difference is that the timing starts to shift. Let us say that, as in Figure 10.2, we are placing a Frame Relay SVC request within an ATM protocol stack. The application (transport layer, in theory) places a NL_CONN_RQ primitive onto the Frame Relay Network Layer. This causes a DL_CORE_DATA_RQ to be sent to the Q.922A module, which then causes a PH_DATA_RQ primitive to be sent to the "physical layer."

Because of the glue logic, the FR_to_ATM function which replaces the PH_ command handler knows that an ATM Virtual Circuit is needed to transfer the data. However, the PH_DATA_RQ primitive will not have the information needed to establish the ATM Virtual Circuit (or to choose the appropriate PVC).

One method of solving this problem is to have the coordinating entity (ITU-T C-plane) have information that maps the physical/logical link being specified by the PH_DATA_RQ to information needed to pick out, or establish, the correct Virtual Circuit. Thus, upon receipt of the NL_CONN_RQ primitive, the FR_to_ATM func-

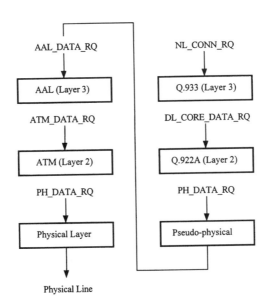

FIGURE 10.2 Primitive flow in a stacked protocol model.

tion can ask the Coordinating Entity (CE) for the appropriate mapping information and status of the Virtual Circuit.

If the VC is already present and active, the information received back from the CE is used only for the appropriate identification information so that the AAL_DATA_RQ primitive can be formatted correctly. If the VC is *not* present or not active, the information can allow the FR_to_ATM module to specify the information necessary for an AAL_CONN_RQ. It waits for an AAL_CONN_CF and then can proceed with the AAL_DATA_RQ primitive.

Another possibility is to make the transport layer intelligent enough about the needs of the different protocol stacks such that it can issue primitives to the appropriate stacks in an order that will satisfy all the protocol needs. In keeping with the above example, the transport layer would receive a request to establish (or gain access to) a Frame Relay VC. It would find that the Frame Relay VC is to be transported across a particular ATM VC. If the ATM VC was not available yet, then layer four would issue the AAL_CONN_RQ primitive and wait for the AAL_CONN_CF before sending the NL_CONN_RQ to the Frame Relay stack.

This option requires a shifting of the primitive flow. With a "normal" glued stack, all PH_ requests and responses going through the glue module are translated into appropriate upper layer primitives of the next stack and all AAL_ indications and confirmations are translated into the appropriate actions for "up-the-stack" transfer. If the transport layer is acting to coordinate the two (or more) stacks, the primitives leading back up must be routed to the Layer 4 module until after the connection is established and then routed to the glue logic module.

Which method is best will depend on the specific host/client architecture in use. The non-coordinated approach (direct replacement) allows the easiest substitution but may affect timer conditions within the "inner" protocol while it is waiting for the "outer" protocol to be ready for its data. This can cause error conditions in the stack and may cause threshing. Use of the coordinated (transport-layer driven) approach eliminates timer situations as the inner layer will not be activated until the outer layer is ready and available. However, it requires a more direct knowledge of the software configuration of the system and the ability to re-route message flows.

10.1.2.3 Data Structure Use

Use of multiple stacks can be confusing, especially if the stacks are not always used together. The ITU-T C-plane and S-plane have some indications as to how they are to be used in conjunction with multiple services but not how they are to be implemented.

One of the ways to do this is to use the concept of *interfaces*. An interface can be a separate physical link (such as multiple lines running from a hub to separate nodes) or it can be a logical interface. An interface corresponds to a protocol stack with its own physical layer. Within the interface, there may be multiple logical channels (which may map to TDM channels or address-differentiated links on the same TDM channel or Virtual Circuits) but all the logical entities will either be part of the same protocol stack or they will be identified according to some known administration protocol. For example, a BRI ISDN has three channels. One channel, the D-channel, is used for signaling, maintenance, and (optionally) multiplexed data protocols such as X.25. The two B-channels are for data use—speech, PPP, V.120, X.25, or whatever is desired. The three channels are grouped as a single interface but the channels have different purposes.

Within the concept of an interface, the various layer protocols can be known, and substituted, as long as they expect to be using the same protocol primitives at a particular layer interface. At the LLD interface, the LLD can be abstracted into function pointers—one pointer for each type of primitive protocol handler (such as one for PH_ primitives, another for MPH_ primitives, etc.). In this way, the layer can be exchanged by just changing the function pointers.

The CE can also maintain data structures that allow the knowledge of the state of each interface (conceptually, from the information sent in the form of management primitives). This allows coordination of different virtual interfaces that are to be used together on a single physical interface.

10.2 SIGNALING

Signaling is a protocol stack, just like any other protocol. However, signaling invokes a *result,* which is the establishment (or failure to establish) a new logical circuit over which data are to be received and transmitted. This means that there must be coordination between the channels. The NL_CONN_RQ primitive will return a NL_CONN_CF that contains information (or verifies information given earlier) about how to access the new logical circuit that has been established.

This can get even more confusing when multiple stacks are being used together. Use of the CE for managing the different signaling and data channel identifiers can make this more manageable. Let us say that NL_CONN_RQ is routed according to the type of service that is requested—BRI signaling, PRI signaling, CAS T1 signaling, Frame Relay Q.933 in-band signaling, X.25 in-band signaling, etc. The CE receives the primitive from the transport layer and then routes it according to need. If it is to be routed to an interface for which there is an "outer" protocol layer, the CE can handle the necessary prerequisites before passing along the primitive.

10.3 STANDARDIZATION

From a technical point of view, industry standardization is not needed for ADSL or any other protocol. As long as all equipment on the particular network—or both ends of a communication line—agree on the protocol (local standardization) the equipment will work well.

From marketing, advertising, sales, deployment, mass production, and ease-of-use points of view industry standardization is very useful. It allows a standard presentation of the product. More direct comparisons of features are possible with competition based on extra features rather than on different architectures. Central Offices and consumers are able to get equipment without having to be concerned with what equipment is located at the other end. Standard firmware and device drivers can be manufactured such that the "core" of the interface devices can be produced in large numbers.

Ease of use is very important for mass deployment. Most consumers don't want to be concerned with just how to install or configure new equipment (we already have enough complications in our lives—we want solutions—not more problems). In the business sector, there is a recognition that added complexity and configuration will cost the company more to make use of the equipment. This is a hidden overhead that must be calculated into the cost.

Cost (more than value), however, is the overriding concern for most consumers and this is the major factor that must be taken into account when deciding upon a standard. This is the subject of great debate within the ADSL news groups and other inter-industry discussion groups. A simple protocol set can be deployed quickly and inexpensively. A too-simple protocol will be a "quick fix" that will end up being discarded as inadequate in a period that is too short to justify use of the device. A complex protocol set will add to the cost of the device and to the difficulty in installing and administering the device. Cost factors will be discussed more in the section on migration needs and strategies.

10.4 REAL-TIME ISSUES

High-speed data protocols deal in very small time units. A 1 Mbps data line potentially receives a byte every 1/125,000 of a second or 125,000 bytes per second. A high-density diskette contains 1.44 Mbytes of data. Thus, a 1 Mbps data line could fill an HD floppy every 8 seconds. It takes a high-end PC about 1 minute to write 1.44 Mbytes to a floppy. Thus, the data line speed, at 1 Mbps, produces data about 7.5 times faster than it is possible to store on floppy disk.

Hard disks are much faster than floppy disks (and RAM access is even faster). Nevertheless, it is reasonable to say that a system using a high-speed access line will need to have a fast hard disk, large amounts of RAM, and a good processor.

10.4.1 BOTTLENECKS

We discussed bottlenecks at the beginning of this book. As we can see above, we had better not be doing an FTP transfer over ADSL to a floppy disk (the above write rates for a floppy disk indicate a transfer time of about 190 kbps)! Every part of the system must be examined in terms of its limitations.

For the present, however, the bottleneck that is likely to be the most difficult to overcome quickly, is the speed of the servers on the net. The majority of ISPs only have a T1 feed into their location. One request for a large file (say 7 Mbytes), will come close to taking over the line. Of course, IP datagram protocols (as discussed in Chapter 8), limit each datagram to 1,500 bytes, which means that what will really happen is that the 1 Mbps transfer time will come in bursts of ,500 byte chunks every once in a while.

The second most difficult bottleneck to overcome will be the Internet itself. As the access speed increases, the total bandwidth demand is likely to continue to

Architectural Issues and Other Concerns   167

increase. The backbones will be overloaded and the smaller network links will collapse.

This doesn't mean that high-speed access mechanisms aren't worthwhile. The parts of the system *will* catch up if there is sufficient demand to make it economically feasible. In the meantime, certain connections will be upgraded first. So, there will be very uneven performance in access. For certain sites that have a high-speed connection for the entire route, with fast servers and high-speed feed lines into the Internet, the speeds of ADSL will be useful. There are also other potential service markets than the Internet, as mentioned throughout the book. If the CO is actively participating in video feeds, the access speeds will be likely to meet the requirements.

10.5 MIGRATION NEEDS AND STRATEGIES

People buy products when they perceive a need for them. At the present, there are two product lines that have significant market penetration and have bandwidth requirements to urge higher speed consumer access. The first is video and audio (including television/cable/satellite broadcasts). The second is the Internet. E-mail is the most-often used service of online service providers and, without large attachments, it does not need broadband capabilities. Other services are useful for niche markets but not significant percentages.

After a need is perceived, the cost of service becomes the deciding issue. If a service that is being promoted as providing Video on Demand (VOD), costs $40/month plus a $2 charge per movie, then (at $1 per movie at low-cost video rental stores) the chance of being able to sell the service to people is virtually nil. The cost must be approximately the same (or less) than what they can get via other methods to bring on significant numbers of clients. Using the aforementioned costs, a fee of $30/month that includes 30 free movie rentals and a $.50/movie rental after that would be competitive.

The Internet is the other strong potential for pushing bandwidth demand (and use for ADSL). However, there is no *requirement* for high-speed to use the Internet. Yes, it is useful but not required. This leads us to the issue of cost versus value. Cost is objective. Value is mostly subjective.

In the VOD example, it can be argued (and will be by sales people selling the service) that the consumer has access to more movies, with easier access and no waiting lines. This is an argument for value provided. Value is a long-term strategy. If you buy a mattress costing $600 that will last 10 years, it is a better value than a mattress costing $400 that will last for 5 years. However, if your budget only allows for $400 then the perceived value for the more expensive mattress is immaterial. People will pay extra for perceived value if they feel a need for the product, can afford it, and cannot get what they would consider to be the equivalent product from some other avenue.

ADSL is being promoted to provide higher-speed customer premises data access on the local loops. However, it is also being considered because it offers an alternative to the providers of the speech network which has not been engineered to support the current data transfer needs.

10.5.1 REPLACEMENT OF LONG-DISTANCE INFRASTRUCTURE

If data network access continues at the existing levels (and every projection indicates that it will be continuing to grow), the long-distance speech networks will have to be upgraded or offloaded. In North America, the cost of use of the speech services for consumers is closely regulated. Furthermore, it would be impossible to determine (without dozens of laws broken) what subscribers are using the network for Internet access and, thus, wrecking havoc with the traffic engineering of the network. Therefore, the providers of speech network services *must* provide alternate ways of funding improvements in the infrastructure.

Raising revenue for infrastructure improvements requires new products that are past basic needs (which are protected closely by the state or provincial regulators). (Note that some of the economic arguments are the same for non-North American areas but government telephone systems shift the direct consumer cost.) Special services such as voice mail, three-way calling, and such are being more actively promoted by the PSTN providers. Other possible services can bring in money for infrastructure improvements.

Selling new products to subsidize infrastructure improvements works but it doesn't meet the network providers real desires: to make a profit on each service. The network providers need to offload the speech network by moving the subscribers who don't meet the (originally) "normal" traffic statistics. BRI and PRI ISDN does not meet this requirement, since the data are still supported by the speech network.

ADSL is basically ideal for this purpose. It uses the same local loop in which the PSTN providers have already invested and the data must go onto a different long-distance network (that can be tariffed in a different, and more profitable manner). Luckily, if the service succeeds, it does *not* have to be sufficiently profitable to subsidize the speech network. This is true because success of ADSL and the alternate long-distance network (ATM cell, Frame Relay, IP router, SONET, or whatever) means that the existing speech network does not have to grow.

It can be argued, therefore, that the operating companies can, and should, provide ADSL (with ISP coverage) for no more than the current cost of an analog line plus ISP service. This is about $40/month. What will happen, however, is unknown.

10.5.2 FTTN, FTTC, AND VDSL

There is nothing tremendously different about Very high-speed Digital Subscriber Line (VDSL). It uses technologies similar to ADSL and offers a basically asymmetric service—at about 10 times the data rate of ADSL. In order to do this, it must do a few things differently. Mainly, it severely limits the distance that it can be from a high-speed feeder. In other words, migration to VDSL is a matter of "bringing the central office closer" rather than trying to make do with speed and distance limitations based on the existing local loop.

Another way of looking at VDSL is from a migration path from the use of Digital Loop Carriers (DLCs). DLCs are being used more and more based on need for new local loops, the need to reuse existing copper for higher bandwidth, and the need for greater bandwidth which can only be addressed by shortening the distance

Architectural Issues and Other Concerns

of the analog local loop. The distance limitations commonly discussed for VDSL are shown in Table 10.1.

VDSL has basically the same set of potential services as ADSL. However, ATM use is at the forefront for one main reason: ATM is currently the best long-distance network format designed to support the speeds of VDSL. The main system envisioned to provide the extension of the CO's link to the local loop is the Synchronous Optical NETwork (SONET).

Support of VDSL is expected to be by use of Fiber To The Neighborhood (FTTN) or Fiber To The Curb (FTTC). FTTN is more likely and it will probably form a hub network as seen in Figure 10.3. SONET uses an ATM cell based layer 2 and layer 3 and this lends greater weight to the argument that ATM should be used on top of VDSL (whether or not this extends to the host systems is another matter).

**TABLE 10.1
Projected VDSL Speeds**

Distance	Downstream Speed	Upstream Speed
4500 feet	12.96 – 13.8 Mbps	1.62 Mbps
3000 feet	25.92 – 27.6 Mbps	3.24 Mbps
1000 feet	51.84 – 55.2 Mbps	6.48 Mbps

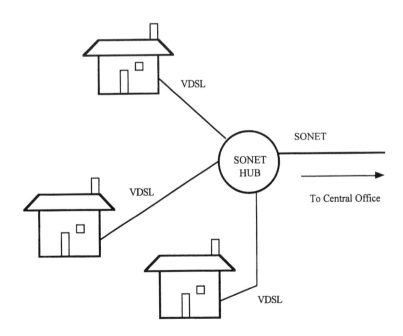

FIGURE 10.3 VDSL with a SONET hub.

10.6 SUMMARY OF ISSUES AND OPTIONS

The existing network was designed to support speech. As a result of the slow building up of this network over the centuries, with speech as the main criteria, many assumptions, as to how the network would be used, were built directly into the design criteria.

First, as the need for greater capacity on the long-distance network grew, the basic unit capacities of the long-distance channels were designed around the needs for speech. Working with Nyquest's Sampling Theorem and a voice bandwidth of about 0 KHz to 3,700 KHz, there was a need for sampling 8,000 times per second (twice the possible sample rate). Using 8-bit codes for each speech sample gave a channel capacity of 64,000 bps. This channel size has permeated digital networks all across the globe.

Second, the network cannot be economically supported to give full access to all users to all other users (full connectivity). This is a matter of traffic engineering. Most of the time, the network is designed such that there is enough capacity for all but a very small percentage of the time. This design was based on a certain number of calls per line per day and a certain average length of a call (about three minutes). Together, this determined just how many "trunk" lines would be supplied and how much capacity the various network switches had to have.

New uses of the speech network have made this traffic model collapse. There are many new services used with the analog speech network—causing a great increase in the number of lines per household (or capita, depending on how you want to list the situation). Even worse, the use of the speech network for data uses has greatly increased the average hold time (the time that the call is active). In a network traffic engineering model, a doubling of the average hold time means that the number of trunk lines needed doubles. This cannot be done immediately and the demands on the network, as data use continues to grow, is threatening to cause breakdowns of the old infrastructure.

Third, various devices have been placed on the local loops to make the quality of speech service better. These devices which improve speech quality do so by focusing on the behavior of the lines in the speech bandwidth. This is to the detriment of the other parts of the potential bandwidth which are needed for additional speed.

There are many different protocols, and line services, available for bringing the digital capabilities of the network to the customer premises. The high end of the MODEM world (the 56K MODEM) still makes use of the speech network. The first fully digital local loop protocols, BRI and PRI ISDN, also make use of the speech networks although they do provide considerably greater data speed. Other xDSL protocols exist by bypassing the speech network and being directed towards IP routers, ATM cell relay switches, Frame Relay networks, and so forth.

A Consumer Digital Subscriber Line (CDSL, also referred to as ADSL "lite") variant of ADSL is regarded as a potential direct replacement for the analog MODEM. This limited version of ADSL allows home installation without the need of technical service people calls. It does this by limiting the options and speed of the data access on the line.

ADSL has had two primary physical line coding techniques used in the experimental arena. These are known as CAP and DMT. In actuality, DMT makes use of CAP technology also. The main difference with DMT is that it splits the frequency spectrum into multiple equal-sized bands which are available for allocation to upstream or downstream traffic. The DMT technology has been chosen as the basis for ADSL standards by ANSI and also the International Telecommunication Union Telecommunications Standardization Sector (ITU-T).

The ADSL physical layer protocol is based on superframes, individual frames, and fast and interleaved data paths. The superframe is broken down into individual frames which hold various control, supervisory, and data information. ADSL "lite" (or CDSL) simplifies the individual frames by eliminating potential configurations and reducing the overhead byte mechanisms by eliminating options.

High-speed data access methods require use of semiconductor chips and particular protocol stacks. Modern data communication protocols make use of the Open Systems Interconnection (OSI) model to split the functions up into separate "layers." The bottom layers are the physical, data link, and network layers and are considered to be the "chained layers" as they exist in some form at each node of a network while the other layers are passed only from end to end.

It is possible to use ADSL as a conduit from the customer premises to the central office, using the local loop as "dry copper" and then routing the data stream directly to a service provider such as an Internet Service Provider (ISP). However, more versatile use can be made of the data link if it has some type of routing, or signaling capabilities. This can be obtained by using some type of protocol on top of ADSL which contains signaling options. Some protocols which contain signaling possibilities include Frame Relay, BRI and PRI ISDN, and ATM.

Asynchronous Transfer Mode (ATM) was designed as a self-contained cell routing mechanism to be used in high-speed cell relay system network switches. It has a high-overhead cell structure but allows sufficient data address information to be able to create high-speed networks. It has a signaling protocol associated with it, called ITU-T Recommendation Q.2931, but many of the existing uses of the ATM network make use of Permanent Virtual Circuits (PVCs) which do not need to have signaling features.

ATM is both a protocol and a layer. The ATM layer occupies the OSI model layer 2 for the data link layer. The ATM Adaptation Layer (AAL) acts as the main layer to be used for data transport. Two AAL Types are recommended for use with ADSL. These are called AAL Type 1 and AAL Type 5. AAL Type 5 is the one recommended for use with ATM over ADSL by the Services Network Architecture Group (SNAG) which operates as a working study group within the ADSL Forum.

As mentioned, ATM has a considerable overhead. For many services, the fast network capabilities of ATM are not needed. Thus, there are also options to use Frame Relay directly over ADSL or use of TCP/IP over ADSL into IP routers. Frame Relay is a highly efficient, streamlined, data protocol that assumes that the physical medium will provide close to an error-free link. TCP/IP is particularly useful because it is used within the Internet. The Internet is one of the primary factors pushing people towards greater network access speeds.

If a system provides high-speed access lines, then the data must be able to be transmitted all the way through to the host device that has requested the data (or which is transmitting the data). This host access link is one of the potential places where a bottleneck can occur. A bottleneck is the spot in a network which can support the lowest data speed. The entire network is limited to the capacity of the slowest part of the network path.

A host access path can be supported by an existing high speed LAN protocol, a new LAN protocol, a high-speed port interface, or some form of data bus extension or direct motherboard support. The major requirement is that it provide a data rate that is at least as fast as the access line.

Note, as mentioned above, that there are many possible places in the network which may place a break on the network speed. The place that is most likely to cause delays is at the server end of an Internet request. Due to relatively low feeder data paths into the servers' locations and possible underdesign of the server computer system, this is the area most likely to cause delays in Internet requests.

ADSL can be a considerable boon to the Public Switched Telephone Network (PSTN) providers as it can offload the existing speech network in favor of use of higher speed data networks. This allows continued use of the existing networks for speech without exceeding the traffic engineering requirements *and* provides a continued growth path for more, and faster, data network access.

Consumers, however, must be persuaded of the need for the speed and must be presented with a cost/benefit situation that is difficult to ignore. In other words, if the direct product can be obtained in some other way (even if not as "well"), ADSL must be presented in such a manner that the cost will be the same or less than possible existing competing services.

Continued increases of access speed to the residences on existing unshielded twisted pairs (UTPs) is unlikely unless the known laws of physics are altered considerably. However, by effectively shortening the distance of the local loops by bringing the high-speed digital networks closer to the ends of the local loop, it is possible to reach much greater speeds than ADSL can achieve. This technology is called Very high speed Digital Subscriber Line (VDSL). It makes use of high-speed feeder hubs that can be gathered from the central office, or central routing area to the neighborhoods. One of the most likely current technologies to be used to provide this hub capability is called Synchronous Optical Network (SONET) but this migration of Fiber-To-The-Curb (FTTC) is likely to extend over the next few decades.

References and Selected Bibliography

The following documents, referred to in the first two sections, are either used as reference material within this book or contain material that may be of interest. The first section refers to ITU-T Recommendations. The next section lists other technical reference specifications. The following section lists Internet Web sites. The final section lists various books on xDSL, ADSL, and ISDN, some of which are referred to in this book. This listing should not be considered to be a complete guide to all xDSL or ISDN literature.

ITU-T RECOMMENDATIONS

F.310	Broadband Videotex Services
F.732	Broadband Videoconference Services
F.821	Broadband TV Distribution Services
F.822	Broadband HDTV Distribution Services
G.703	Physical/Electrical Characteristics of Hierarchical Digital Interfaces
G.704	Synchronous Frame Structures Used at Primary and Secondary Hierarchical Levels
G.711	Pulse Code Modulation (PCM) of Voice Frequencies
G.721	32 kbit/s Adaptive Differential Pulse Code Modulation (ADPCM)
G.722	7 kHz Audio Coding Within 64 kbit/s
G.922.1	(G.dmt) Asymmetrical Digital Subscriber Line (ADSL) Transceiver
G.922.2	(G.lite) Splitterless Asymmetrical Digital Subscriber Line (ADSL) Transceivers
G.994.1	(G.hs) Handshake Procedures for Digital Subscriber Line (DSL) Transceivers
G.995.1	(G.ref) Overview of Digital Subscriber Line (DSL) Recommendations
G.996.1	(G.test) Test Procedures for Digital Subscriber Line (DSL) Transceivers
G.997.1	(G.ploam) Physical layer management for Digital Subscriber Line (DSL) Transceivers
I.112	Vocabulary of Terms for ISDNs
I.113	Vocabulary of Terms for Broadband Aspects of ISDN
I.120	Integrated Services Digital Networks
I.121	Broadband Aspects of ISDN

I.122	Framework for Providing Additional Packet-Mode Bearer Services
I.140	Attribute Technique for the Characterization of Telecommunication Services Supported by an ISDN and Network Capabilities of ISDN
I.150	B-ISDN ATM Functional Characteristics
I.211	B-ISDN Service Aspects
I.230	Definition of Bearer Service Categories
I.231	Circuit-Mode Bearer Service Categories
I.233	Frame Mode Bearer Services
I.311	B-ISDN General Network Aspects
I.320	ISDN Protocol Reference Model
I.321	B-ISDN Protocol Reference Model and its Application
I.325	Reference Configurations for ISDN Connection Types
I.327	B-ISDN Functional Architecture
I.361	B-ISDN ATM Layer Specification
I.362	B-ISDN ATM Adaptation Layer (AAL) Functional Description
I.363	B-ISDN ATM Adaptation Layer (AAL) Specification
I.411	ISDN User-Network Interfaces—Reference Configurations
I.413	B-ISDN User-Network Interface
I.430	Basic User-Network Interface—Layer 1 Specification
I.431	Primary Rate User-Network Interface—Layer 1 Specification
I.432	B-ISDN User-Network Interface—Physical Layer Specification
I.440	*See* Q.920
I.441	*See* Q.921
I.450	*See* Q.930
I.451	*See* Q.931
I.452	*See* Q.932
I.460	Multiplexing, Rate Adaptation, and Support of Existing Interfaces
I.610	OAM Principles of B-ISDN Access
Q.2931	Broadband Integrated Services Digital Network (B-ISDN) — Digital Subcriber Signalling System No. 2 (DSS 2) —User Network Interface (UNI) Layer 3 Specification for Basic Call/Connection Control
Q.920	ISDN User-Network Interface Data Link Layer—General Aspects
Q.921	ISDN User-Network Interface Data Link Layer Specification
Q.922	ISDN Data Link Layer Specification for Frame Mode Bearer Services
Q.930	ISDN User-Network Interface Layer 3—General Aspects
Q.931	ISDN User-Network Interface Layer 3 Specification for Basic Call Control
Q.933	Digital Subscriber Systems No. 1 (DSS 1) Signalling Specification for Frame Mode Basic Call Control
V.42	Error-correcting Procedures for DCEs Using Asynchronous-to-Synchronous Conversion

OTHER TECHNICAL REFERENCES

ANSI T1.413	Network and Customer Installation Interfaces—Asymmetric Digital Subscriber Line (ADSL) Metallic Interface. Issue 1,1995. Draft Issue 2, December 1998.
RFC 791	Internet Protocol
RFC 793	Transmission Control Protocol
RFC 1490	Multiprotocol Interconnect over Frame Relay
RFC 2400	Internet Official Protocol Standards
RFC 2460	Internet Protocol, Version 6 (IPv6) Specification

SELECTED INTERNET UNIFORM RESOURCE LOCATORS (URLs)

http://www.adsl.com	ADSL Forum home page
http://www.ansi.org	ANSI home page
http://www.atmforum.org	ATM Forum home page
http://www.etsi.fr	ETSI home page
http://www.frforum.com	Frame Relay Forum home page
http://www.ieee.org	IEEE home page
http://www.ietf.org	Internet Engineering Task Force
http://www.itu.int	ITU home page
http://www.usb.org	Universal Serial Bus home page

SELECTED BIBLIOGRAPHY

Bell Laboratories, *Engineering and Operations in the Bell System,* Bell Telephone Laboratories, 1977.

Carlson, James, *PPP Design and Debugging*, Addison-Wesley, Reading, MA, 1997.

Chen, Walter Y., *DSL Simulation Techniques and Standards Development for Digital Subscriber Line Systems*, Macmillan Technical Publishing, Indianapolis, IN, 1998.

Comer, Douglas E., *Interworking With TCP/IP: Principles, Protocols, and Architechture*, Prentice Hall, NJ, 1995.

Ferrero, Alexis, *The Evolving Ethernet*, Addison-Wesley, New York, 1996.

Goralski, Walter, *ADSL and DSL Technologies*, McGraw-Hill, New York, 1998.

Händel, Rainer, and Manfred N. Huber, *Integrated Broadband Networks: An Introduction to ATM-Based Networks*. Addison-Wesley, Reading, MA, 1991.

Kessler, Gary C., and Southwick, Peter V., *ISDN: Concepts, Facilities, and Services*, Signature Edition, McGraw-Hill, New York, 1998.

Lishin, Pete, *TCP/IP Clearly Explained*, 2nd edition., Academic Press Professional, Boston, MA, 1997.

Rauschmayer, Dennis J., *ADSL/VDSL Principles: A Practical and Precise Study of Asymmetric Digital Subscriber Lines and Very High-Speed Digital Subscriber Lines*, Macmillan Technical Publishing, Indianapolis, IN, 1999.

Simoneau, Paul, *Hands-On TCP/IP*, McGraw-Hill, New York, 1997.

Stallings, William, *ISDN and Broadband ISDN with Frame Relay and ATM*, 4th ed., Prentice-Hall, Inc., NJ, 1998.

Stallings, William, *Networking Standards: a Guide to OSI, ISDN, LAN, and MAN Standards*, Addison-Wesley, Reading, MA, 1993.

Summers, Charles K., *ISDN Implementor's Guide: Standards, Protocols, and Services*, McGraw-Hill, New York, 1995.

INDEX

A

AAL, *See* ATM Adaptation Layer
AAL_ primitives, 120
Access fees, 28
Access line capacity, 18
Acknowledgment, proprietary protocol requirements, 145–146
Actions, 91, 121
Adaptive asynchronous protocols, 37
ADB, 157
Address fields, 97, 141
   destination, 97, 103, 141
   Ethernet frame, 152–153
   Frame Relay, 105, 131–132
   IP, 103–104
   origin, 98, 103, 141
   TCP virtual circuits ("port addresses"), 142
ADSL (Asymmetric Digital Subscriber Line), xix, 23, *See also* Digital Subscriber Line (xDSL) technologies; ISDN
   access unit "class" definition, 43
   hardware access, *See* Hardware access and interactions
   host access, *See* Host access
   industry standardization, 165, *See* ADSL standardization
   ISDN architecture and, 23–24, *See also* ISDN
   physical layer protocol, xx, 24–25, 47–66, *See* Physical layer
   POTS ports, 32, 43
   programming interface, 60
   protocol stacks, *See* Protocol stacks
   splitterless system, *See* ADSL "lite"
ADSL Forum, 21, 23, 25, 26, 28, 43, 126
ADSL interface chip, 74–75
ADSL "lite," 30, 44–45, 62–64, 170
ADSL/RADSL, 29, 42–43
ADSL standardization, 25–27, 159, 165, 171, *See also* ANSI; ITU-T; specific protocols
Always On/Dynamic ISDN (AO/DI), 37
American National Standards Institute, *See* ANSI
American Wire Gauge (AWG) standard, 6
Amplification, 16
Analog data communication, 21–22
Analog devices, signaling methods, 98–101
Analog Interfaces (AI), 74
Analog-to-digital conversion, 22
ANSI, 25, 26, 171
   Frame Relay recommendations (T1.606, T1.617, T1.618), 130
   T1, 38–41
   T1.413, 26–27, 43, 50–61, 107
   T1.606, 130
   T1.617, 130–131
   T1.618, 131
   T1E1.4, 42
ANSI/IEEE 802.3, 149
Anycast, 104
Apple Desktop Bus (ADB), 157
Application access, 79–80, *See* Host access
Application layer, 71, 72
Application Programming Interface (API), 94
Arbitration, 83
Area codes, 99
Asymmetric Digital Subscriber Line, *See* ADSL
Asynchronous adaptive protocols, 37
Asynchronous messages, 88–89
ATM (Asynchronous Transfer Mode), xxi, 23, 42, 54–55, 107–127, 171
   complexity, 107
   data organization (PDUs), 108
   distribution services, 109
   fast and interleaved data, 57
   fast byte/sync byte, 58
   hardware support, 81
   history, 108
   interactive services, 109
   ITU-T Recommendations, 109, *See* ITU-T
   market considerations, 107
   multi-protocol stacks and, 160–163
   Network Timing Reference (NTR), 111
   OSI layers, 110
   physical layer, 111, 146, *See* Physical layer
   primitives, 120
   signaling, 120–126
      B-ISDN message set, 123–124
      default codeset, 126
      disconnection and release request, 122
      general architecture, 120–121
      Information Elements, 125–126
      lower layer access, 120
      network-side states, 122–123
      user-side states, 121–122
   User-to-Network and Network-to-Network Interfaces, 113
   virtual circuits, 66, 163–164
   virtual paths and channels, 115–116
ATM Adaptation Layer (AAL), 81, 110, 113, 116–119, 171
   AAL_ primitives, 120
   protocol types table, 119
   Segmentation and Reassembly (SAR), 81, 116–118

Type 1, 117, 118, 171
Type 5, 117, 118, 171
  multi-protocol stacks and, 161
ATM cells, 104–105, 111
  formats, 113–115
ATM Forum, 25, 26
ATM Layer, 104, 110, 111–116, *See also* Data link layer
ATM25, xxi
ATM25 interface, 147

B

Backplane architectures, 157
Backward Explicit Congestion Notification (BECN), 133
Bandwidth requirements for human speech signal transmission, 14–15
Basic Rate Interface Integrated Services Digital Network, *See* BRI-ISDN
B-channels, 35, 108–109, 165, *See also* Bearer channels
Bearer channels, 51, 54–55, 108–109, *See also* B-channels
Bell Laboratories, 99
Binary coding, 3
B-ISDN, *See* Broadband ISDN
Bottlenecks, 17–19, 166–167, 172
BRI-ISDN, xx, 21, 29, 31–32, 170, *See also* ISDN
  cost, 28
  data protocols, 37
  DSLAM and WAN access, 66
  IDSL, 37–38
  multi-protocol stacks and, 160–161
  physical layer, 32–35
  primitives, 72, 102
  signaling protocol, 102
  switching protocol, 35–37
  Terminal Adaptor, 32
  transport classes, 55
Bridged taps, 15, 16
Broadband ISDN (B-ISDN), xxi, 23, *See also* ATM
  bearer services, 108–109
  history, 108
  ITU-T Recommendations (F.2xx series), 109
  message set, 123–124
  OSI layers, 110
  physical layer, 111
  signaling, 120–126, *See under* ATM
  User-to-Network and Network-to-Network Interfaces, 113
Broadcast media, 7
Buffer, 11, 79

Buffer descriptors, 86–87
Bus control, 83

C

Cable modems, xix, 18
Call Reference Value (CRV), 36, 123
Carrier Sense Multiple Access/Collision Detection (CSMA/CD), 150
Carrierless Amplitude/Phase (CAP) modulation, 42, 47–49, 171
CCITT, 23, *See* ITU-T
CDSL, 30, 44–45, 50, 170, *See* ADSL "lite"
Cell Loss Priority (CLP), 113
Cell relay, xxi, *See* ATM
Chained layers, 71, 89
Channel Associated Signaling (CAS), 41, 101
Checksum, 144–145
Circuit, 8, 11
Circuit-switching, 9–11
Clock rate, 4
CODEC translations, 75
Command housing, 88
Command registers, 85
Command/Response (C/R) bit, 132
Common Channel Interoffice Signaling (CCIS), 101
Communication forms, 1–5
Confirmation primitive, 72, 102, 120
Congestion control, Frame Relay system, 133–134
Connectionless (CL) services, 117
Connectivity, 98
Constant Bit Rate (CBR) services, 117
Constellation encoding, 74
Consumer Digital Subscriber Line (CDSL), 30, 44–45, 50, 170, *See* ADSL "lite"
Control fields, 134, 146
Control plane, 73
Control systems, 80
Convergence Sublayer (CS), 81, 116–117
Copper wiring transmission media, xix, xx, 6
  managing signal degradation and attenuation, 15–17
CopperGold ADSL transceiver, 95
Coprocessor systems, 77, 147
Cosine wave, 47–49
Cost issues, 28, 166–168
Cyclic Redundancy Check (CRC), 36, 60, 70, 163–164
  AAL Type 5, 118
  Ethernet frame, 153
  fast byte use, 58
  physical layer semiconductors, 75

Index

D

Data acknowledgment, proprietary protocol requirements, 145–146
Data buses, 79
Data bus extension, 157
Datagrams, 139–141, 166
Data identification, proprietary protocol requirements, 145
Data integrity, proprietary protocol requirements, 144–145
Data Link Connection Identifier (DLCI), 105, 132
Data link core primitives, 134–136
Data link layer, 69–70, 101, 104, 138, 171, *See also* ATM Layer
   Frame Relay, 130–131
   IP frame, 139
   multi-protocol stacks and physical layer replacement, 162
   proprietary protocol requirements, 146
   xDSL (BRI ISDN), 37
Data recovery, proprietary protocol requirements, 146
Data transfer protocols, 125, *See* Frame Relay; Internet Protocol; Transmission Control Protocol
D-channel, 35, 165, *See also* LAPD
Destination address field, 97, 103, 141
Device driver library, 94–95
Dial tone, 100
Digital data communication, 21–22
Digital Interfaces (DI0, 74
Digital Loop Carriers (DLCs), 15, 16–17
   migration issues, 168–169
Digital modems, 18, 22
Digital Signal Processing (DSP), 74
Digital Subscriber Line (xDSL) technologies, xix, 21–46; *See also* ADSL; ADSL "lite"; ATM; BRI-ISDN; 56K MODEM; ISDN; specific protocols, standards bodies, technologies
   ADSL/RADSL, 29, 42–43
   CDSL/ADSL Lite, 44–45
   cost, 28
   data protocols, 37
   family of protocols, 28–46
   HDSL/HDSL2, 38–41
   physical layer, 32–35
   rate adaptive variants, 18
   table of technologies and features, 29–30
   terminology, 21
   VDSL, 28, 30, 45
Digital-to-analog conversion, 75
Digital transmission coding, 3–5
Discard Eligibility (DE) bit, 133
Disconnect signal, 122
Discrete MultiTone (DMT), 42, 49–50, 107, 171
Disk I/O transfer time, 17–18
Distance limitations, 15, 42
Distribution services, 109
DL_primitive, 72
DLCI, 105, 132
DMT, 42–43, 49–50, 107, 171
DSLAM (DSL Access Module), 19, 43, 64–66, 105
Dual Tone MultiFrequency (DTMF), 100–101
Dumb board application, 76
Duplex bearers, 55
"Dying gasp" signal, 61

E

E1 system, 39, 40
   transport classes, 54
Echo cancellation, 49
Embedded Operations Control (EOC), 58, 60–62
   ADSL "lite" (splitterless ADSL), 63
   device driver library, 95
Encapsulation, 79, 102, 103, 162
EOC, *See* Embedded Operations Control
Ephemeral ports, 142
Erasable Programmable Read-Only Memory (EPROM), 81
Error correction, *See also* Cyclic Redundancy Check; Embedded Operations Control
   CRC and HDLC framing, 70
   data link protocol, 69
   Forward Error Correction (fec), 57
   Frame Relay protocols, 131
   proprietary protocol requirements, 146
Ethernet, 70, xxi, 127, 138
   Fast Ethernet, 147
   frame, 152–153
   history, 149
   MA bridges, 154
   Medium Access Control (MAC), 149, 150–151
   OSI model equivalents, 149
   physical medium and protocols, 154
European Telecommunication Standards Institute (ETSI), 25, 27
Events, 83, 90–91, 121
Extended ISA, 157

F

Fast byte, 58
Fast data, 57
Fast Ethernet, 147

FDM, 12–13
Fiber optics, 7
Fiber-To-The-Curb (FTTC), 2, 45, 169, 172
Fiber To The Neighborhood (FTTN), 169
FIFOs, 85–86
56K MODEM, xx, 21–22, 28, 29, 31, 170
File Transmission Protocol (FTP), 72, 166
First-In; First-Out queue (FIFO), 85–86
Flash memory, 81
Floppy disks, 166
Flow-control, 79
Forward Error Correction (fec), 57
Forward Explicit Congestion Notification (FECN), 133
Frame Check Sequence (FCS), 36, 131, 153
Frame delineation, 70
Frame-Mode Connection Control Messages, 136
Frame-oriented protocols, 37, *See* Frame Relay; ML-PPP
Frame Relay, xxi, 129–138, 171
    congestion control, 133–134
    control field, 134
    data link core primitives, 134-136
    data link layer, 130–131
    DSLAM connection, 66
    error recovery protocols, 131
    history, 130
    link access protocol, 131–134
    multiple protocol considerations, 137–138, 160, 162–163
    Network Layer signaling, 136
    protocol set, 105
    virtual circuits, 129
Frame Relay Forum, 105, 130, 131
FRF.4, 136
Frame Relaying Specific Convergence Sublayer (FRCS), 118
Frame Relay Network, 133
Frequency Division Multiplexing (FDM), 12–13
FRF.4, 136

G

Gateways, 71
Generic Flow Control (GFC), 113, 116
"Glue logic," 82, 163
Glue module, 164
"G."-series of ITU-T Recommendations, *See under* ITU-T

H

Handshake, 75, 143
Hard disks, 166

Hardware access and interactions, xxi, 81–96, 171
    ADSL chipset interface example, 94–96
    data bus extension, 157
    direct microprocessor access, 147–148, 157–158
    interface chip, 74–75
    Low-Level Drivers (LLDs), 87–89
    microprocessor direct access, 158
    physical layer semiconductors, 75
    state machines, 89–94
    synchronous and asynchronous messages, 88–89
    system configuration design, 75–76
    Universal Serial Bus (USB), 147, 155–157
Hardware components and interactions, 73–77, *See also* Hardware access and interactions
Hardware/software interface registers, 84
Hardware timer, 82–83
HDLC, *See* High-level Data Link Control; LAPD
HDSL, 21, 29, 38–42
HDTV, 109
Header Error Control (HEC), 70, 112, 116
Header fields, 139–140, 142–144
High bit-rate Digital Subscriber Line (HDSL), 21, 29, 38–42
HDSL2, 32, 42
High-level Data Link Control (HDLC), 36, 69–70, *See also* LAPD
    error checking, 70
    error correction, 60
    Ethernet frame and, 153
    multi-protocol stacks and, 160
    physical layer semiconductors, 75
High-pass filter, 51
Host access, xxi, 79–80, 147–158, 172
    direct microprocessor access, 147–148, 157–158
    Ethernet, 147, 148–155
    motherboard support, 157–158
    Universal Serial Bus (USB), 147, 155–157
    "upper layers" model, 71
Host-controlled systems, 76–77
Host I/O capacity, 17–18
HyperText Markup Language (HTML), 72, 109

I

IDSL, 29, 37–38
IEEE
    Ethernet standard (802.3), 149, 154
    Sub-Network Access Protocol (SNAP), 138
IETF RFCs, 137
In-band signaling, 101

Index

Indication primitive, 72, 102, 120
Indicator bits, 58, 60
Industry Standard Architecture (ISA), 157
Industry standardization, 165–166, *See* ADSL standardization
Information Elements (IEs), 125–126, 136
Infrastructure limits, 13–17
Initialization commands, 88, 95
Integrated Services Digital Network, *See* ISDN
Interactive services, ATM over ADSL, 109
Interface chip, 74–75
Interfaces, multi-protocol stack implementation context, 164–165
Interlayer primitives, 72, 102
 Frame Relay data link core primitives, 134–136
Interleaved data, 57
International Organization for Standardization (ISO), 67
International Telecommunication Union-Telecommunication Standardization Sector, *See* ITU-T
Internet
 access fee, 28
 bottlenecks, 166–167
 pushing bandwidth demand, 167
Internet Assigned Numbers Authority (IANA), 142
Internet Engineering Task Force (IETF) RFCs, 137
Internet Protocol (IP), xxi, 97, 138–141
 datagrams, 139–141, 166
 "header fields," 139–141
 IPv6, 103–104, 140, 142
 network layer protocol, 70–71
 routing and addresses, 103–104
Internet Service Provider (ISP) signaling protocol, 78
Internetworking, 35, *See* Interworking
Interrupt control groups, 83
Interrupt mask, 88
Interrupt Service Routine (ISR), 88
Interrupt servicing, 88
Interworking
 B-ISDN state specifications, 121
 OSI model, 35
 protocol stack considerations, 77, 78–79
Intranet systems, 79, 127, 138
I/O port transfer rate design criteria, 18
I/O requests, 79, 84
IP, *See* Internet Protocol
IPv6, 103–104, 140, 142
ISA, 157
ISDN (Integrated Services Digital Network), 23, *See also* BRI-ISDN; Broadband ISDN

 ADSL and, 23–24
 basic architectural model, 24
 global connectivity, 32
 plane architecture, 110
 signaling using Primary Rate Interface ISDN, 41
ISDN Digital Subscriber Line (IDSL), 29, 37–38
ISO, 67
ISO/DIS 88302-3, 149
ITU-T, 23, 171
 B-ISDN (F.2xx series), 109
 G.922.2, 62–64, *See* ADSL "lite"
 G.944.1, 95
 G.992.1 (G.dmt), 27
 G.992.2 (G.lite), 27, 95
 G.994.1 (G.hs), 27
 G.996.1 (G.test), 27
 G.997.1 (G.ploam), 27
 I.113, 108
 I.121, 108
 I.122, 130, *See also* Frame relay
 I.211, 108–109
 I.361, 114
 I.362, 116
 I.370, 134
 I.430, 68
 I.432, 111
 I.610, 114
 Q.921, 35–36, 42, 66, 101–102, 131, *See* LAPD
 Q.922, 105, 131, 134–136
 Q.922A (Annex), 105, 130, 131, 134–136, *See also* Frame Relay
 Q.931, 35, 42, 36, 66, 101–102, 130
 Q.933, 105, 130
  Frame-Mode Connection Control Messages, 136
 Q.2931, 120–126
 V.90, 29, 31, *See* 56K MODEM
 V.110, 37
 V.120, 37, 130
 V.120, 37
 X.20 bis, 74, 75
 X.25, 37, 71, 101
 X.31, 129, *See also* Frame Relay

K

K56Flex, 31

L

LAPB, 101
LAPD, 36, 101, 105

LAPF, 102, 130
LAPM, 21, 102
Layer 2 Tunnelling Protocol (L2TP), 138
Line conditioning, 31–32
Link Access Procedures, *See* LAPB; LAPD; LAPF; LAPM
Link access protocol for Frame Relay, 131–134
Loading coils, 15–16
    line conditioning, 31–32
Local Area Network (LAN), 11
    DSLAM, 64
Local loop distance limitations, 15
Local Management Interface (LMI), 131
Long-distance capacity, 19
Long-distance trunk lines, 9
    digital system, 22
    DLCs, 17
    infrastructure replacement, 168
    multiplexing, 13
Low-Level Drivers (LLDs), xxi, 69, 87–89
    device driver library, 94–95
    interface, 165
    multi-protocol stacks and, 161
Low-pass filter, 51

M

Machine bus, 157
Mailbox system, 161
Management Information Base (MIB) group, 26
Management layer primitives, 72
Management plane, 110
MCA, 157
Medium Access Control (MAC), 149, 150–151, 154
Memory mapping, 80, 84
Metasignaling, 116, 120, 126
Micro Channel Architecture (MCA), 157
Microwave transmission, 7
Migration needs and strategies, 167–169
ML-PPP, 37, 108
Modems, xix, xx, 21, 100
    cable modems, xix, 18
    56K MODEM, xx, 21–22, 28, 29, 31, 170
    host I/O capacity and, 18
    Link Access Protocol, 21
    protocols, 37
Morse code, 2, 4
Motherboard support, xxi, 157–158
MPEG, 47
MPEG-II, 109, 129
Multicast, 104
Multi-Link Point-to-Point (ML-PPP) protocol, 37, 108

Multiplexing, 12–13
Multiprotocol Interconnect over Frame Relay, 137
Multi-protocol stacks, 160–165
    architectural choices, 160–161
    coordination tasks, 163–164
    data structure use, 164–165
    physical layer replacement, 162–163
    signaling, 165
    software implementation, 161–162
    virtual circuits and, 163–164

N

Narrowband-ISDN, 108
Network-to-Network Interface (NNI), 113
Network administration applications, 77
Network Facility Associated Signaling (NFAS), 41
Network layer, 70–71, 101–102
    B-ISDN signaling protocol, 120
    Packet Protocol, 102
    signaling, for Frame Relay, 136
Network Layer Protocol IDentifier (NLPID), 138, 160
Network saturation, 19
Network Timing Reference (NTR), 51, 111
N-ISDN, 108
NL primitive, 72
Nyquest's Sampling theorem, 4, 14, 170

O

OHCI, 155
Open Systems Interconnection (OSI) model, 67–73, 171, *See also* Protocol stacks; specific layers
    B-ISDN (ATM) protocol model, 110
    BRI-ISDN protocols, 35–37
    Ethernet and, 149
    interlayer primitives, 72, 102
    layer 1, 68–69, *See* Physical layer
    layer 2, 69–70, *See* Data link layer
    layer 3, 70–71, *See* Network layer
    layer 4, 71, *See* Transport layer
    older data protocols and, 70–71
    protocol hierarchies, 126–127
    protocol modularity, 73
    upper layers, 71–72
Operating systems, support for asynchronous messages, 89
Organizationally Unique Identifier (OUI), 138
Origination address field, 98, 103, 141

Index

OSI, *See* Open Systems Interconnection (OSI) model
OSIG support, 121
Overlap receiving (U25), 121
Overlap sending (U2), 121
Overrun condition, 86

P

Packet-switching, 9–11
PCI, 157
Peripheral Component Interconnect (PCI), 157
Peripheral plug-and-play capability, 155
Permanent virtual circuits (PVCs), 66, 97, 104, 105, 129, 171
Personal Communication Systems (PCS), 7
PH_primitive, 72
Physical driver, 69
Physical interface chip, 81
Physical layer, xx, 47–66, 101, 171
  ANSI T1.413, 50–61
  ATM, 111
  ATM25 interface, 147
  bearer channels, 51, 54–55
  BRI ISDN architecture, 32
  Carrierless Amplitude/Quadrature Amplitude Modulation (CAP/QAM), 47–49
  CRC bits, 60
  Discrete MultiTone (DMT), 42, 49–50
  DSLAM components, 64–66
  embedded operations control, 58, 60–62
  fast byte/sync byte, 58
  fast data, 57
  indicator bits, 58, 60
  interleaved data, 57
  multi-protocol stacks and, 160
  protocol stack architecture (OSI model), 68–69
  receiver-central unit (ATU-R ATU-C) matching, 64
  replacement and adding protocol stacks, 162
  semiconductors, 75, *See also* Semiconductor devices
  specifications, 24–25
  superframe structure, 55–60, 171
  synchronization or "activation," 69
  transport classes, 54
Physical Medium (PM), 69, 111
  Ethernet, 154
Pin-outs, 81–83, *See also* Hardware access; Semiconductor devices
Plain Old Telephone Service (POTS), xix, *See* Public-switched telephone network
POTS ports, 32, 43
Plane management functions, 110
Plug-and-play capability, 155
Point-to-Point Protocol (PPP), 66, 127, 139, 160
Polling, 83, 88
Port addresses, 142
Power-up sequence, 88
Presentation layer, 71, 72
PRI ISDN, 29, 41, 170
Primary Rate Interface ISDN (PRI ISDN), 29, 41, 170
Primitive interfaces, 88
Private Branch Exchanges (PBXs), 121
Private Integrated services Network eXchanges (PINXs), 121
Proprietary protocol design requirements, 129, 144
  data acknowledgment, 145–146
  data identification, 145
  data integrity, 144–145
  data protocol, 146
  data recovery, 146
Protocol Data Units (PDUs), 108–109
Protocol Discriminator, 123
Protocol identifiers, 138
Protocol modularity, 73
Protocol stacks, 77–79, 171
  interworking, 77, 78–79
  multi-protocol stacks, 160–165
    architectural choices, 160–161
    coordination tasks, 163–164
    data structure use, 164–165
    physical layer replacement, 162–163
    signaling, 165
    software implementation, 161–162
  OSI model, 67–73, *See* Open Systems Interconnection (OSI) model
  signaling, 77, 78
  stack combinations, 79
Protocol state machines, 89–94
Pseudoternary method, 33
Public-switched telephone network (PSTN), 98–101
  ADSL and infrastructure replacement, 172
  BRI ISDN Terminal Adaptor, 32
  CDSL/ADSL "lite,", 43–45
  challenge of high-speed and multiple-access demand, xx
  DSLAM, 65–66
  long-distance infrastructure replacement, 168
  switch hierarchy, 99
  telephone number system, 98–99
  POTS ports, 32, 43
  speech-based design criteria, xx, 14–15, 170

Pulse Code Modulation (PCM), 31, 78

Q

"Q."-series of ITU-T Recommendations, *See under* ITU-T
Quadrature Amplitude Modulation (QAM), 42, 47–49

R

Race condition, 88
Radio wave transmission, 7
RADSL, 29, 42–43, 45
Random Access Memory (RAM), 81
  buffer descriptors, 86–87
  host I/O capacity and, 17
Rate-adaptive ADSL (RADSL), 29, 42–43, 45
RBOCs, 28, 38
Read-Only Memory (ROM), 81
Real-time issues, 166–167
Real-time tasking system, 161
Regional Bell Operating Companies (RBOCs), 28, 38
Registers, 84–85
Release request signal, 122
Repeaters, 16
Request primitive, 72, 102, 120
Response primitive, 72, 102, 120
Retrieval services, 109
RLCG parameters, 15
Robbed bit signaling, 41, 101
Route, 8
Routers and routing methods, 11–12, 77, 97, 102–105
  ADSL and ISDN architecture, 23–24
  ADSL/RADSL signaling, 42
  ATM cells, 104–105
  DSLAM, 65–66, 105
  Frame Relay, 105
  Internet Protocol, 103–104
  permanent virtual circuits, 104
  security checks, 103
Routing tables, 102

S

Sampling rate, 4–5, 14, 170
SCSI, 157
SDSL, 29, 32, 42, 47
Security issues, 101, 103
Segmentation and Reassembly (SAR), 81, 116–118
Semaphore method, 161

Semiconductor devices, xxi, 73, 81, 171
  access, 81–87, *See* Hardware access and interactions
  Low-Level Drivers and, 87–89
  physical layer and, 75
Sequence Number (SN), 117
Sequence Number Protection (SNP), 117–118
Serial Line IP (SLIP), 139
Serial ports
  transfer rate design criteria, 18
  UART chip interface, 75
Server access line and performance, 17
Service Access Point (SAP), 116
Service Access Point Identifier (SAPI), 105
Services Network Architecture Group (SNAG), 158, 160, 171
Service-Specific Connection-Oriented Protocol (SSCOP), 118
Session layer, 71, 72
SHDSL, 42
Signal degradation and attenuation
  distance limitations on local loops, 15
  loading coils and, 15–16
  repeaters and line extenders and, 16
Signaling, 97–102
  analog and digital signals, 2
  ATM, 120–126, *See under* ATM
  Channel Associated Signaling (CAS), 41, 101
  defined, 97
  DSLAM, 105
  Link Access protocols, 101–102
  methods, 98–102
    analog devices, 98–101
    DTMF, 100–101
    Q.921/Q.931 variants, 101–102
  multiple protocol stacks, 165
  network layer signaling for Frame Relay, 136
  PRI ISDN, 41
  protocol stack considerations, 77, 78
  state machine based, 97
Simple and Efficient Adaptation Layer (SEAL), 118
Simplex bearers, 54
Sine wave, 47–49
Single-pair Digital Subscriber Line (SDSL), 29, 32, 42, 47
Single-pair High-bit-rate Digital Subscriber Line (SHDSL), 42
Small Computer Systems Interface (SCSI), 157
SNAG, 158, 160, 171
Software interface LLDs, *See* Low-Level Drivers
Software timer, 83
SONET, 111, 169, 172
Specification Definition Language (SDL), 92

Index

Speech-based telecommunications technology and design criteria, xx, 14–15, 170
Splitterless ADSL, 62–64, *See* ADSL "lite"
Standalone systems, 77
Standards, 24–27, 159, 165–166, 171, *See also* ADSL Forum; ADSL standardization; ANSI; ITU-T
Starting Frame Delimiter (SFD), 152
State event tables, 77
State machines, 89–94, 120
   implementation methods, 92
   specifications, 91–92
   traffic light control system example, 92–94
States, 89–90
State table, 91, 120, 126
STM, *See* Synchronous Transfer Mode
STM-1 bit interface, 111–112
Streaming data, 129
Sub-Network Access Protocol (SNAP), 138
Superframe structure, 55–60
Switched Virtual Circuits (SVCs), 66, 105, 129, 130–131
Switching, 8–11
   BRI-ISDN protocols, 35–37
Sync byte, 58
Synchronization frame, 55
Synchronization handshake, 75, 143
Synchronous messages, 88–89
Synchronous Optical Network (SONET), 111, 169, 172
Synchronous Transfer Mode (STM), 50, 51
   ATM CELLS, 111–112
   fast byte/sync byte, 58
   STM-1 bit interface, 111–112
System configuration design, hardware considerations, 75–76

T

T1, 38–41
T1.413, 26–27, 107
T1.606, 130
T1.617, 130–131
T1.618, 131
T1E1 standards, 26–27
Tasking system, 89, 161
TDM, 12–13, 35
Telephone number system, 98–99
Telephone system, existing technology and capacity, xix–xx, *See also* Public-switched telephone network
Terminal Adaptor (TA), 22
Terminal Endpoint Identifier (TEI), 105
Terminal Equipment (TE), 22

Three-dimensional coding schemes, 5, 47
Time Division Multiplexing (TDM), 12–13
Timer leads, 82–83
Timer primitives, 90
Time-to-Live field, 140
Token network, 150–151
Traffic engineering, 9
   Frame Relay congestion control, 133
Traffic light control system example, 92–94
Transmission code, 4
Transmission Convergence (TC) sublayer, 69, 70, 111, 149
Transmission media, 5–7
Transport classes, 54
Transport Control Protocol (TCP, TCP/IP), xxi, 66, 71, 127, 129, 138, 141–144, 171, *See also* Internet Protocol
   header fields, 142–144
   multi-protocol stacks and, 160
   virtual circuits, 142
   features, 144
Transport layer, 71
   Multiprotocol Interconnect over Frame Relay, 137
Tunnelling protocol, 79, 127, 129, 138

U

UART chip, 75
UHCI, 155
UI field, 134
Underrun condition, 86
Unicast, 104
Universal ADSL Working Group (UAWG), 25, 26
Universal Asynchronous Receiver Transmitter (UART) chip, 75
Universal Host Controller Interface (UHCI), 155
Universal Serial Bus (USB), xxi, 147, 155–157
Unnumbered Information (UI) field, 134
Upper layers, 71–72
USB (Universal Serial Bus), xxi, 147, 155–157
User Datagram Protocol (UDP), 141
User plane (U-plane), 110
User-to-Network Interface (UNI), 113
UTP, 49

V

VDSL, *See* Very high-speed Digital Subscriber Line
Versa Module Eurocard (VME), 157
Very high-speed Digital Subscriber Line (VDSL), 28, 30, 45, 168–169, 172
   distance limitations, 169

migration from digital loop carriers, 168–169
 optical fiber network support, 169
Very large-scale integrated (VLSI) circuit, 47
Video applications, 109, 129
Video on Demand (VOD), 47, 57, 167
Virtual Channel Connections (VPCs), 115
Virtual Channel Identifier (VCI), 104, 115–116
Virtual circuits, 66, 97, 129, 171
 multi-protocol stacks and, 163–164
 Permanent virtual circuits (PVCs), 66, 97, 104, 105, 129, 171
 Switched Virtual Circuits (SVCs), 66, 105, 129, 130–131
 TCP/IP, 142
Virtual Path Connection Identifier (VPCI), 120
Virtual Path Connections (VPCs), 115–116
Virtual Path Identifier (VPI), 104, 115–116, 120
VME, 157

"V."-series of ITU-T Recommendations, *See under* ITU-T

W

Wide Area Network (WAN), 12
 DSLAM connection, 64, 66
Windows, 144
Wire gauge, 6, 15

X

xDSL, *See* Digital Subscriber Line (xDSL) technologies
XMODEM, 71

Z

Zero-bit insertion, 70